DIGITAL TIMING MEASUREMENTS

FRONTIERS IN ELECTRONIC TESTING

Consulting Editor
Vishwani D. Agrawal

Books in the series:

Fault-Tolerance Techniques for SRAM-based FPGAs
Kastensmidt, F.L., Carro, L. (et al.), Vol. 32
ISBN: 0-387-31068-1
Data Mining and Diagnosing IC Fails
Huisman, L.M., Vol. 31
ISBN: 0-387-24993-1
Fault Diagnosis of Analog Integrated Circuits
Kabisatpathy, P., Barua, A. (et al.), Vol. 30
ISBN: 0-387-25742-X
Introduction to Advanced System-on-Chip Test Design and Optimi...
Larsson, E., Vol. 29
ISBN: 1-4020-3207-2
Embedded Processor-Based Self-Test
Gizopoulos, D. (et al.), Vol. 28
ISBN: 1-4020-2785-0
Advances in Electronic Testing
Gizopoulos, D. (et al.), Vol. 27
ISBN: 0-387-29408-2
Testing Static Random Access Memories
Hamdioui, S., Vol. 26
ISBN: 1-4020-7752-1
Verification By Error Modeling
Redecka, K. and Zilic, Vol. 25
ISBN: 1-4020-7652-5
Elements of STIL: Principles and Applications of IEEE Std. 1450
Maston, G., Taylor, T. (et al.), Vol. 24
ISBN: 1-4020-7637-1
Fault injection Techniques and Tools for Embedded systems Reliability...
Benso, A., Prinetto, P. (Eds.), Vol. 23
ISBN: 1-4020-7589-8
Power-Constrained Testing of VLSI Circuits
Nicolici, N., Al-Hashimi, B.M., Vol. 22B
ISBN: 1-4020-7235-X
High Performance Memory Memory Testing
Adams, R. Dean, Vol. 22A
ISBN: 1-4020-7255-4
SOC (System-on-a-Chip) Testing for Plug and Play Test Automation
Chakrabarty, K. (Ed.), Vol. 21
ISBN: 1-4020-7205-8
Test Resource Partitioning for System-on-a-Chip
Chakrabarty, K., Iyengar & Chandra (et al.), Vol. 20
ISBN: 1-4020-7119-1
A Designers' Guide to Built-in-Self-Test
Stroud, C., Vol. 19
ISBN: 1-4020-7050-0
Boundary-Scan Interconnect Diagnosis
de Sousa, J., Cheung, P.Y.K., Vol. 18
ISBN: 0-7923-7314-6
Essentials of Electronic Testing for Digital, Memory, and Mixed Signal VLSI Circuits
Bushnell, M.L., Agrawal, V.D., Vol. 17
ISBN: 0-7923-7991-8
Analog and Mixed-Signal Boundary-Scan: A guide to the IEEE 1149.4 Test...
Osseiran, A. (Ed.), Vol. 16
ISBN: 0-7923-8686-8

DIGITAL TIMING MEASUREMENTS
FROM SCOPES AND PROBES TO TIMING AND JITTER

by

Wolfgang Maichen
Teradyne Inc.
Agoura Hills,
California, USA.

Springer

A C.I.P. Catalogue record for this book is available from the Library of Congress.

ISBN-10 0-387-31418-0 (HB)
ISBN-13 978-0-387-31418-1 (HB)
ISBN-10 0-387-31419-9 (e-book)
ISBN-13 978-0-387-31419-8 (e-book)

Published by Springer,
P.O. Box 17, 3300 AA Dordrecht, The Netherlands.

www.springer.com

Printed on acid-free paper

Printed in the Netherlands.

Dedication

This book is dedicated to

my wife Gwen

and

my son Alexandre.

Contents

Preface

As many circuits and applications now enter the Gigahertz frequency arena, accurate digital timing and jitter measurements have become crucial in the design, verification, characterization, and application of electronic circuits. To be successful in this endeavor, an engineer needs a knowledge base covering instrumentation, measurement techniques, jitter and timing concepts, statistics, and transmission line theory. Very often even the most experienced digital test engineer – while mastering some of those subjects – lacks systematic knowledge or experience in the high speed signal area.

The goal of this book is to give a compact yet in-depth overview on all those subjects. The emphasis is on practical concepts and real-life guidelines that can be readily put into practice. It unites in one place a variety of information relevant to high speed testing and measurement, signal fidelity, and instrumentation.

To keep the text easily readable – after all this is an introductory book – we refrained from putting references directly into the text. For readers interested in further study of any of the subjects we highly recommend checking out the "References and Further Reading" section at the end of this book which contains the titles of a large selection of additional books, papers, and application notes that were the basis for this work.

A final word to all numerical values – for reasons of consistency this book uses international (SI) units throughout – this means that some numerical factors in the formulas are different compared to the original literature they were taken from.

We always appreciate feedback. If you have comments or questions, please direct them to **Wolfgang@TestTechniques.com**.

Chapter #1

ELECTRICAL BASICS

1. TIME DOMAIN AND FREQUENCY DOMAIN

1.1 What Is The Frequency Domain, Anyway?

When we look around us in daily life, things are changing – their properties are different at different instants in time. E.g. the outside temperature changes during the course of the day and over the year. The same is true when we look at electronic circuits carrying signals – things like the voltage at a given circuit point or the current through a resistor change over time. So what is "real" is the time domain – properties changing over time, and at any given instant they have a well defined value.

Nevertheless, sometimes engineers and scientists are using a different "domain", the frequency domain. There must be good reasons for that, but at first glance this new domain seems rather strange and unintuitive. Let's have a closer look, concentrating on the application to electrical signals, and also let's keep the mathematical apparatus to a minimum!

1.2 Moving Between Domains

1.2.1 The Discrete Fourier Transformation

In time domain the so-called "independent parameter" is – not too surprising – the time. The dependent parameter may be e.g. the voltage on the output of a driver circuit, and it changes in dependence of the time. We could for example plot this dependency (change) in the usual x-y plot. Now mathematics tells us that if the signal is periodic with a period of T, then we can think of it as consisting of a sum of sine waves of different frequencies f_k, amplitudes a_k, and phases φ_k. Mathematically, if the signal is given as a function $s(t)$ over time, then we can *decompose* it into sinusoidal frequency components as follows:

$$s(t) = a_0 + \sum_{k=1}^{\infty} a_k \times \sin(2\pi \times f_k \times t + \varphi_k)$$
$$f_1 = \frac{1}{T}, \text{ and } f_k = k \times f_1 \quad (1)$$

The first coefficient (a_0) is also called the *DC offset* and gives the average value over the period. The slowest sine wave (frequency f_1) is the *fundamental*. It corresponds to the period of the signal (e.g. if the signal repeats every second, then the fundamental is 1/1 sec = 1 Hz, and the higher components have frequencies of 2 Hz, 3 Hz, and so on). The higher frequency components are called “harmonics“ (f_2 is the *second harmonic*, f_3 is the *third harmonic* and so on), and their frequencies are integer multiples of the fundamental.

The mathematical algorithm for determining the coefficients a_k, f_k, and φ_k from the time dependent signal is called “Fourier Transformation“. Since we are unlikely to ever do those very tedious numerical calculations by hand, we will refrain from giving the exact formulas – let’s just accept that our computer or oscilloscope can perform it fast and automated – there exists a very efficient (= fast) though rather unintuitive numerical implementation called “Fast Fourier Transformation” (FFT) that is virtually always used in practical applications.

Often the amplitude and the phase of each component are lumped together into a single complex number

$$c_k = (a_k \times \cos(\varphi_k) + i \times a_k \times \sin(\varphi_k)), \quad (2)$$

but this is a purely mathematical trick that makes certain operations more compact, but that does not change the information content at all. We should note, however, that we can only reconstruct the original waveform if we have *both* the amplitudes as well as the phases of all the Fourier components. This is illustrated in Figure 1(a) and (b): While in both cases the same two sine waves are added up, the difference in phases between Figure 1(a) and (b) results in completely different signals.[1]

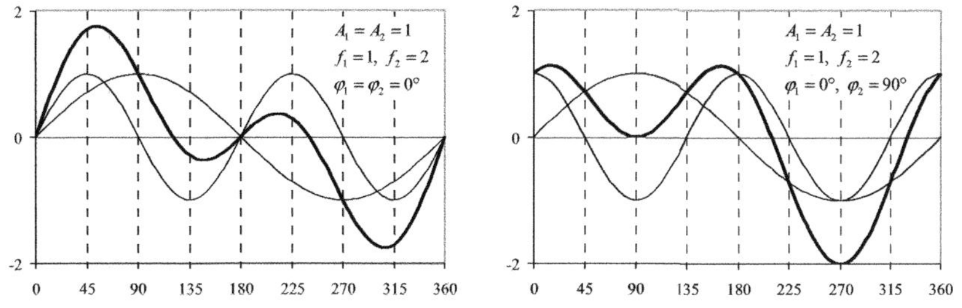

Figure 1: Depending in their relative alignment (i.e. their phase) combinations of the same sine waves can result in very different final results.

To illustrate the Fourier Transformation, let's take a square wave of period T. The fundamental frequency f_1 is simply $1/T$. Decomposing the signal for this particular waveform yields only the odd components with decreasing amplitude (all even components a_{2k} are zero), and the phases are all zero if we assume that the square wave starts at zero at time zero:

$$a_{2k+1} = \frac{2}{\pi \times (2k+1)}, \quad \varphi_k = 0 \quad (k \geq 0)\,. \tag{3}$$

If we sum up more and more of them, the resulting waveform approximates the original square wave better and better, as is shown in Figure 2. In theory we would need an infinite number of components to reproduce the waveform exactly, but in practice it is usually sufficient to keep the first five of six at best, or as few as the first three if we are willing to make some concessions to waveform fidelity: We can see that the components become weaker with increasing frequency, so above a certain limit they have only negligible influence on the wave shape. This is the real reason behind the requirement that a filter must have a bandwidth of at least 3 times (better 5 to 6 times) the signal bandwidth of the incoming signal in order to pass the signal undistorted.

[1] Another, more mathematical way to see this is that the "real" Fourier transformation yields *complex* amplitudes while the SA only displays their absolute values.

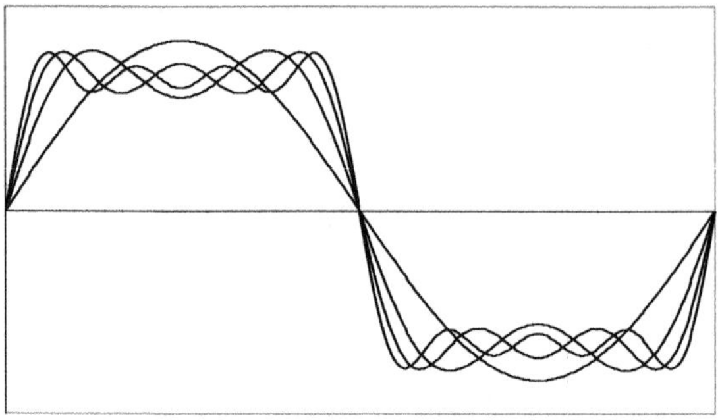

Figure 2: The more harmonics are added up, the closer the waveform gets to its final shape (in this case a square wave).

The components can be displayed graphically as a *discrete frequency spectrum*, displayed for the above square wave in Figure 3. The height of each line corresponds to the amplitude of each component, and its horizontal position indicates its frequency. This is the picture we would also get from a spectrum analyzer (see later). Note that this neglects the phases of the components, so in general this plot alone does not contain enough information to reconstruct the waveform. It does however give us a good idea of the necessary bandwidth to transmit the signal.

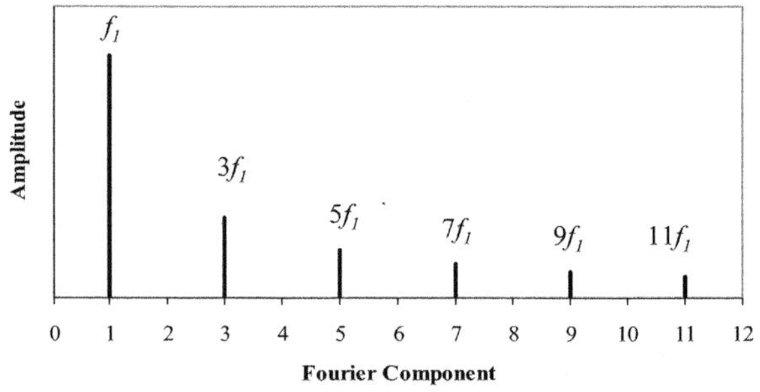

Figure 3: Fourier spectrum of a square wave. The fundamental (f_1) is the strongest component, higher harmonics become smaller and smaller.

Using the signal $s(t)$ or instead the representation as a sum of sine waves is absolutely equivalent (we should mention that if we want to be mathematically rigorous, this is only true whenever the function $s(t)$ is continuous, i.e. has no sudden jumps, but in the real world there are no infinitely fast changes anyway). In other words, so far the decomposition has been a purely mathematical exercise. But we can already see how the name frequency domain and time domain correspond: In one we have the signal amplitude vs. the time; in the other we have the sine wave amplitudes vs. the frequency. We should also note that sine waves are not the only possible "building blocks" to decompose a waveform into, but they are the most widely used.

1.2.2 Linear Systems

Things become immediately more interesting when the circuit the signal is going into is a *linear system*. Without going into theoretical details, a linear system is any system where the size of the *response* to some *stimulus* is linearly proportional to the stimulus. An ideal resistor is a good example: If we apply twice as high a voltage (the "stimulus"), the current (the "response") will double as well. One important result of this is that the *response to the sum* of two stimuli is equal to the *sum of the responses* to each stimulus alone. Taking again the example of the resistor, if we apply a 1 V signal to a 2 Ω resistor, a current of 0.5 A will flow. If we apply a 4 V signal, the current will be 2 A. If we apply now (1+4) = 5 V, and the resistor behaves in a linear fashion (no heating or the like which would change its resistance), then the current will be (0.5+2) = 2.5 A.

Let's apply this to the Fourier decomposition of our signal. What we want to know is how the system will react to our stimulus. If the system is linear, then we can instead first calculate how it reacts to each frequency component alone, and then sum up all those results. Now the reaction of some electronic component to a simple sine wave is usually much easier to calculate and to measure than the response to some arbitrary signal, so this exercise can save us some serious effort. Consider for example an ideal inductor – its impedance for a sine wave is proportional to the frequency, and the phase is a constant -90 degrees. In general, for not too large signals (to avoid heating effects), almost all passive networks consisting of capacitors, inductors, resistors, and many amplifiers – as long as they operate well below saturation – behave as linear circuits. Most of our test equipment – cables, oscilloscopes, spectrum analyzers etc. – is designed to behave linear as well. And sine waves are pretty hardy beasts – if we send a sine wave into a system, chances are we will get out something that will still closely resemble a sine wave, albeit with modified amplitude and phase. The same cannot be said of most other wave shapes.

Having said that, there are still plenty of systems that are not linear. Diodes or saturated transistors are prime examples, as are resistors when the current gets large enough to cause significant heating. If we take a normal diode, then it will "open up" only above the forward bias voltage of around 0.7 Volt. So if we apply one signal of 0.5 Volts amplitude, and then later another signal of same size and frequency, nothing will come out – the effect of each of the two stimuli alone is zero. But if we apply both together (giving a wave of 1 Volt amplitude), we will have significant current – the response to the sum here is definitely different from the sum of the responses!

1.2.3 Aperiodic Signals

If the signal period gets longer and longer, the fundamental – being the inverse of the period – becomes smaller and smaller, and all the harmonics move closer together (since they are spaced from each other by the fundamental). Note that a long period does not mean that the signal has a low bandwidth. It can still have very fast changes (e.g. steep rising and falling edges). All it means is that the signal *repeats* only after a long period of time.

In the limit of an infinite period – in more everyday-terms: it repeats never, it is an aperiodic signal – the discrete Fourier frequency spectrum becomes a continuous spectrum. While this case is very important in mathematics, it has little application in the real world because first we can never measure for an infinite time, and we haven't got infinite timing resolution either, so all our data sets will consist of only a finite number of data points. Nevertheless we often *approximate* a dense discrete spectrum with a continuous spectrum.

Given that our data sets always cover only a limited time span, we don't really care what happens outside this span as far as our data processing is concerned. It does not make a difference to us if the signal repeats after this interval (like a clock signal) or if it is aperiodic (e.g. an isolated pulse that is zero before and after). So we usually go with the easier choice and assume it repeats, allowing for a simple discrete Fourier analysis where the period is equal to the length of our data set.

1.3 Digital Data Streams

In traditional RF applications we usually deal with signals that are very close to sine waves, and small aberrations thereof.[2] E.g. we may have a strong radio carrier signal at some frequency, that is then for example amplitude modulated, creating two equally spaced sidebands, one on either side of the carrier (see Figure 4). In this case looking at the signal in the frequency domain makes a lot of sense, because the picture is simpler than in the time domain, which is why this domain is so popular with RF engineers.

[2] For the uninitiated, "RF" stands for "radio frequency" – pointing back to a time where the main application for high-frequency technology was radio transmission.

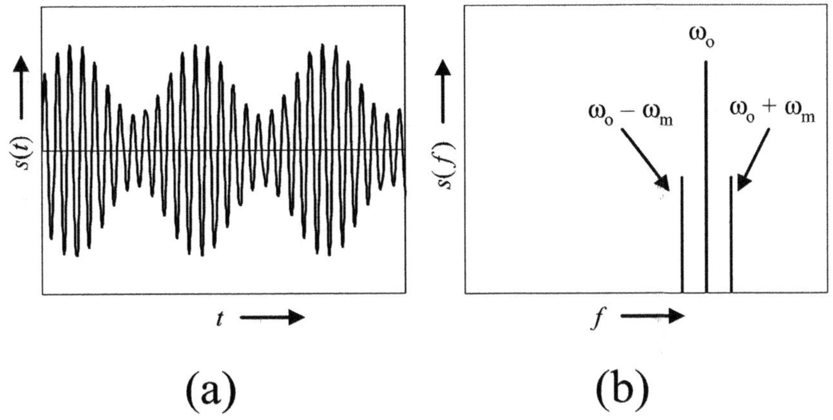

Figure 4: An amplitude modulated sine wave signal (a) in the time domain, and (b) in the frequency domain.

In the digital world things are quite different. Signals look more square-like, not sinusoidal. They are usually not repetitive (except for clock signals), but contain streams of ones and zeroes of more or less random lengths. They may do nothing for a while and then suddenly have a few transitions, then do nothing again.

Seen in the time domain, this behavior has nothing odd to it – we simply see the data bits that get driven. But the corresponding frequency spectrum is hopelessly complicated: Sequences of single ones followed by single zeroes correspond to a fundamental of half the data rate, sequences of more bits of the same polarity (e.g. several ones in a row followed by zeroes) correspond to integer fractions of this frequency. And since the edges (transitions) are not sinusoidal in shape, they correspond to a continuous spectrum of higher frequencies.[3] Figure 5 shows a pseudo-random bit stream (in non-return format) in both the time domain as well as in the frequency domain. There is really no hope of just looking at the frequency spectrum and figuring out what the corresponding bit pattern looks like, even though it is a rather simple one. At least the bit frequency is visible as a series of zeros (dips) in the frequency spectrum[4].

[3] As we will see in the next section, except for pure sine waves the highest bandwidth is not given by the clock frequency or data rate, but instead by the signal rise time.

[4] At first glance it may be surprising that a non-return-formatted bit stream of data rate f has *zeros* (and not *maxima*) at multiples of the data rate. The mystery is quickly solved, however, when one remembers that the maximum frequency clock that can be driven by such a bit stream is only half the data rate – it takes two transitions, one rising and one falling, for a full cycle. And a square wave consists of only odd harmonics, the even harmonics (corresponding to multiples of the data rate) are zero. Any asymmetry in the

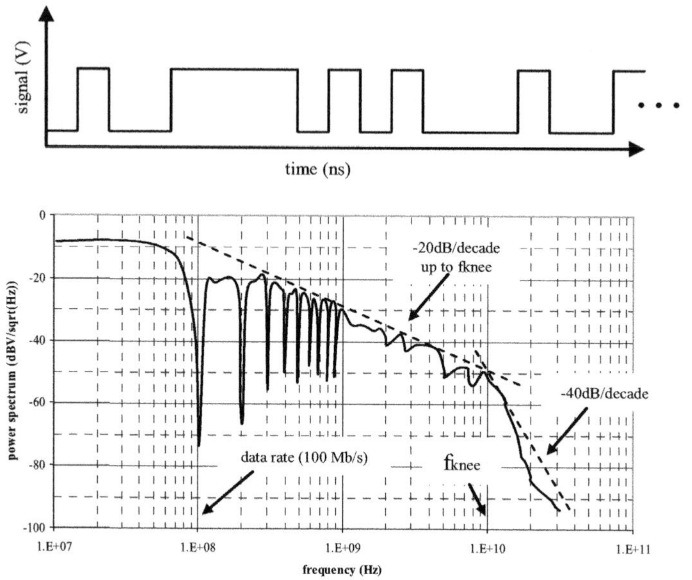

Figure 5: Digital data stream in the time domain (a) and in the frequency domain (b).

Figure 5 also illustrates the main difference between RF type signals and digital signal: While RF signals are usually narrow-band signals (i.e. a particular signal covers only a frequency range that is very small compared to its absolute frequency), digital signals are inherently wide-band, covering the range from fractions of the data rate (dependent on the maximum run length, i.e. then longest stream of all-zeros or all-ones that occurs[5]) up to well beyond the bit rate (dependent on the rise time, as we will see in the following section). That's why the frequency domain makes so much sense for RF signals (they tend to look simple and nicely bounded there). It is much less helpful for analyzing digital data streams where we always have to deal with a full, wide spectrum at once, while in the time domain it is rather easy to separate one bit from the next.

Overall, frequency domain methods widely employed in RF test and measurement applications have only limited appeal for the digital test

waveform (e.g. difference between rise and fall time, or between overshoot and undershoot) will wash out the zeros, however.

[5] However, modern serial transmission schemes often encode the data in a way to limit the maximum run length a just a few bits (e.g. 5 in case of 8b/10b (8bit/10bit) encoding) to allow the use of AC coupling (see section 5.2.4) and clock recovery (see section 8.2.1.3), and to reduce data dependent jitter (see section 8.3.7). This limits the low-frequency content of the signal and reduces the total bandwidth requirement on the low side (the spectrum drops off to zero below a certain limit, e.g. at one fifth of the fundamental for 8b/10b encoding. The fundamental in GHz is of course half the data rate in Gb/s).

engineer dealing with time domain measurements, unless we deal with periodic (clock-like) signals. That said, they can sometimes give valuable additional insight into signal behavior, especially for periodic effects that lend themselves to Fourier decomposition. For troubleshooting, the Fourier transformation into the frequency domain really excels at highlighting even the smallest periodic effect that would be impossible to discern in time domain. Later in this book we will learn about such techniques like phase jitter and periodic jitter measurements. After all, as we now know frequency and timing are closely related through the Fourier transform, so at least in principle it is always possible to acquire a measurement value in one domain and then transfer it into the other. And it never hurts to peek over the fence (in our case between digital and RF) and see if the other side has developed some interesting methods or techniques that one could "steal".

1.4 Signal Bandwidth

One very common mistake that engineers make when they want to decide if a signal is "high-speed" is to look only at the system clock frequency. Of course "high-speed" is a rather vague term, but one good definition for our purposes could be that a high-speed signal is anything that behaves significantly different from DC, and/or that needs leading-edge equipment to be transmitted and measured[6]. Clearly that implies that the frontier between low-speed and high-speed is constantly advancing as technology is making progress.

So why is looking at only the clock frequency such a bad idea? It is because it neglects the rise time of the signal. Figure 6 shows two clock signals that have the same frequency, but very different rise times. Obviously the rate of change of the second one is much higher, and this corresponds to a larger portion of high-frequency components in the frequency domain[7]. If we are to transmit (or measure) those signals without too much distortion from their original shape, the one with the faster rise times clearly needs higher-bandwidth (i.e. faster, in terms of analog performance) equipment, independently of the number of transitions in a given time interval (i.e. their frequency). *Clock frequency is not a good indicator!*

What is the necessary bandwidth BW to transmit a signal with a given rise time T_R? Obviously, since time and frequency are inverse parameters, we expect some behavior along the lines of

[6] Another possibility is to define "high-speed" as the range where the transmission path no longer looks lumped, but becomes a transmission line – more to that in the next section.

[7] Just remember how the edges of the square wave got steeper the more harmonics (with higher and higher frequency) we added!

$$BW = \frac{k}{T_R}. \tag{4}$$

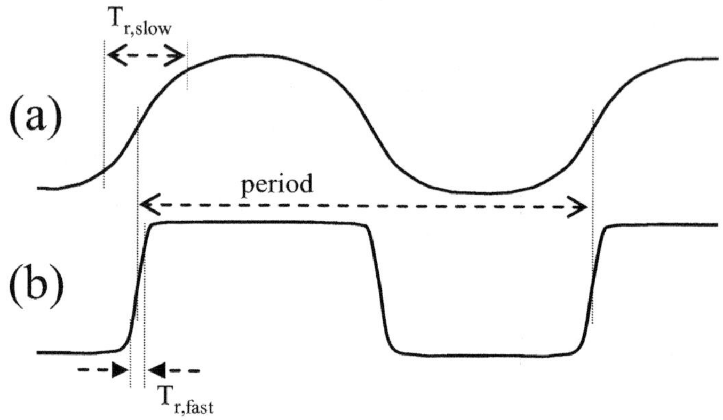

Figure 6: Two signals with same frequency (or data rate), but different rise times. Signal (a) has a slower rise time than signal (b) and thus a lower bandwidth.

As we will see in the next section (talking about low-pass filters), if we choose the bandwidth to be the -3 dB bandwidth, the factor k becomes something around 0.33 for smooth, Gaussian edges. But -3 dB means an attenuation of only 30%, so there will still be a sizeable portion of higher-frequency components beyond that bandwidth left over. Fortunately, if we turn our attention back to Figure 5 which shows a data stream in the frequency domain, we see that above a certain frequency the spectrum tends to drop of very fast. This frequency is called the *knee frequency* and for typical digital signal edge shapes is not too far beyond the -3 dB bandwidth, namely around

$$f_{knee} = \frac{0.5}{T_R}. \tag{5}$$

If in our measurement we include all frequency components up to f_{knee}, signal distortion will be negligible[8].

[8] For linear ramps (trapezoidal waveforms), the drop-off over frequency is -20 dB per decade up to the knee frequency, and -40 dB per decade beyond the knee frequency. The power content beyond the knee frequency is only a tiny fraction of the total power, so omitting (filtering) it will not visibly change the waveform. For electromagnetic emission, however, it still contributes a sizeable amount, so engineers concerned with EMI tend to be interested in the content up to several times the knee frequency.

1.4.1.1 Transmission Line Theory

After just reading the title of this section, our first instinctive reaction could be "All I want to learn about right now is timing *measurements*, not circuit design, so why would I bother with passive circuit theory, signal propagation, and signal integrity"?

Well, as it turns out there are two good reasons. On one hand every measurement setup needs to get the signals from the system under test to the measurement device. This inevitably involves things like cables, connectors, printed circuit boards (PCBs), maybe some filters, etc. And at the speeds we are working with today the simple picture of e.g. a cable as "a connection from point A to point B" with some ohmic resistance or regarding it as a simple capacitive load (as done very often for slower-speed scope probes) just does not cut it anymore. Rise times and clock periods are by now much shorter than the propagation time through the typical cable (and often even the typical connector), so even smallest parasitic imperfections (capacitances, inductances) will have time constants that can distort our signals and produce serious reflections. If we want to stand any realistic chance of really measuring our system under test and not just the weaknesses and flaws in our test setup, we must have at least a working knowledge of transmission lines, parasitics, loss effects, and so on.

On the other hand, as we will see in the chapter about jitter and jitter analysis, many jitter components stem from transmission line effects – reflections, losses, parasitic low-pass filters, and crosstalk to name just a few. In this case knowledge of those topics will enable us to find the root cause of the jitter we measured, and to improve either our test setup or the system under test.

1.5 Low-pass Filters

When one hears of filters, one instinctively thinks of rather complex networks of active and/or passive components. But in the area of signal measurement setups and of signal integrity, the term "low-pass filter" is mostly used to express that a cable, probe, connectors or some other part of our setup cannot transmit arbitrarily fast changes in the signal (the time domain view) or that it dampens higher frequencies stronger than lower ones (the frequency domain view). It is clear that we will rarely put such filters into our setup on purpose, but that they come about because of technological, physical, or budget reasons that force us to make do with equipment and components that hopefully fulfill their purpose, but possibly not much more, and which is never ideal.

1.5.1 Rise Times

There are several different possibilities to define the rise time (or the fall time) of a signal transition. The intuitive way would be to use the time it takes for the signal to rise from the start level to the final level, but in practice this definition meets serious problems: First, what is the final level for the common case that the signal settles rather slowly? Is it the level attained after 1 ns, or after 100 ns, or after one hour? Second, every signal has at least some small noise, which again makes it impossible to define an exact low or high level. So we need a more practical definition.

A very common definition is the time it takes the signal to go from some intermediate level, e.g. 10% along the full swing, to some other such level, e.g. 90%. This is indicated in Figure 7(a). At those levels the signal is rising nicely, so a small uncertainty in the actual threshold levels causes only negligible uncertainty in the corresponding timing instants. We call the rise time thus defined the 10/90 rise time $T_{10/90}$. If the signal settles very slowly, is very noisy, or the swing is small, we may want to use the 20% and the 80% level instead, yielding the 20/80 rise time $T_{20/80}$, respectively. Those are the rise time definitions we will use throughout the book.

Further possibilities to indicate rise times would be the center rise time (rise time of the tangent to the steepest part of the edge) or the maximum slew rate (level change per unit of time, e.g. in V/ns).

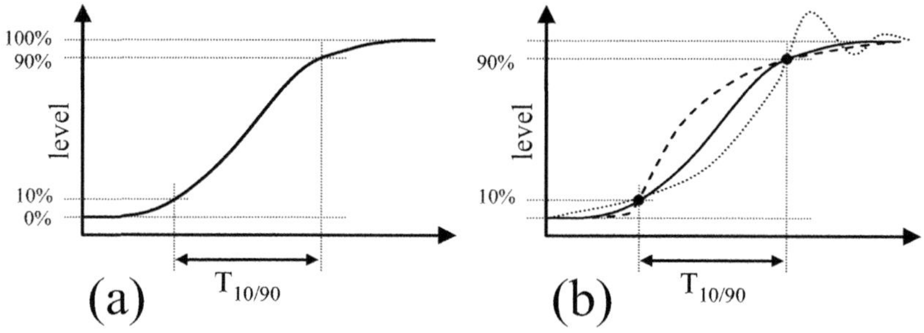

Figure 7: (a) Definition of the 10/90 rise time $T_{10/90}$. (b) Very different signals can nevertheless have the same rise time.

The advantage of $T_{10/90}$ (or $T_{20/80}$) is that it is very simple to measure manually, and in addition most modern oscilloscopes can determine it automatically. It is also the parameter given in virtually any device or instrument specification. On the other hand it depends on only two points of the waveform and does not take into account the behavior of the waveform at any other instant. It should be clear that this inevitably means some loss of information – in general one just cannot stuff the information of a two-dimensional curve into a limited set of values. Turning this view around, the fact that two signals have the same rise time does not mean they will look or behave exactly the same. For example, as shown in Figure 7(b), very different edge shapes can produce the same $T_{10/90}$. But in many practical cases their behavior will be sufficiently similar to merit those simplifications.

1.5.2 Filter Bandwidth, Time Constant, and Rise Time

As we indicated in the beginning, we have always the option of looking at a signal waveform or a filter response in either the time domain or in the frequency domain. The latter is very common and well established from the RF world, but is unfortunately a bit unintuitive for us dealing with digital signals, where the time domain reigns.

In principle the response (i.e. what comes out) of a filter to an incoming signal can be very complex. But to make things easier to grasp and to be able to compare different filters, people try to reduce this complexity to just a few numbers. In time domain a very fundamental filter parameter is the so-called filter rise time $T_{r,filter}$. This is the rise time of the output signal if the input signal is rising infinitely fast (the so-called "step response" of the filter). The definition of the filter rise time is thus analogous to the rise time of a signal. Another time domain parameter is the so-called time constant T_c. This parameter is mainly applicable to simple 1-pole filters like an R-C filter ($T_c = R \times C$) or an R-L filter ($T_c = L/R$), and it is the time from the start of the edge to (1-1/e) $\approx$ 63% of the full swing. Its connection with $T_{10/90}$ is simple:

$$T_{10/90} \approx 2.2 \times T_c \,. \tag{6}$$

Since historically low-pass filters were mostly a concern in the RF world, in many cases what is specified (e.g. for a cable, connector, oscilloscope or other element in our setup) is the -3 dB bandwidth BW_{-3dB}. This is the frequency where the filter attenuates the signal by -3 dB, or in other words the output signal has only 70% of the input signal's amplitude.[9] Higher frequencies will be attenuated even more strongly. Time and frequency are inversely proportional to another, so it comes to no surprise that this is also reflected in the connection between filter bandwidth and filter rise time (or, for that matter, signal bandwidth and signal rise time):

[9] Since we are in the frequency domain, those signals we are talking about are always sine waves! For general signals we need to decompose the signal into its Fourier spectrum and treat each frequency component separately.

$$BW_{-3dB} = \frac{k}{T_{10/90}}. \qquad (7)$$

The exact factor k depends on the signal (or filter response) shape:

$$\begin{array}{ll} \text{Gaussian edge:} & k = 0.33 \\ \text{Exponential edge:} & k = 0.35 \\ \text{Modern digital oscilloscopes:} & k \approx 0.4 \\ \text{Butterworth filter:} & k = 0.49 \\ \text{Chebyshev filter:} & k = 0.6 \end{array} \qquad (8)$$

For the case of $T_{20/80}$ the formula stays the same, except for different k factors (for Gaussian edges, $k \approx 0.22$). The last two filter types (Butterworth and Chebyshev) are mostly used in RF applications. They are less useful in time domain because their output signal shows strong overshoot and ringing. We can also see that the factor for "well-behaved" signal (or filter) shapes does not change very much with different edge shapes. Each of the filter types has some specific advantages. The most important properties of each of above filter types are:

- Butterworth filter: Has a very flat response (loss) curve in the pass band, and then falls off with -20dB/decade/pole[10] above the -3 dB point. *In time domain it exhibits strong overshoot and ringing.* Its group delay[11] is not constant vs. frequency (and thus time delay vs. signal rise time). This

[10] All practical, real-world filters are made up by a limited number of stages. Casually speaking, the more stages there are the closer the filter will be to its ideal behavior, and the steeper the potential drop-off above the -3 dB bandwidth can be. A simple R-C element is an example for a one-pole filter, with -6dB/octave (-20dB/decade) drop-off in gain above its bandwidth. Two R-C sections with identical bandwidths in series create a two-pole filter with -12dB/octave drop-off. Typical commercial filter designs use up to over 10 poles.

[11] In the frequency domain elements can be characterized by their gain (ratio between output amplitude and input amplitude).and their phase (relative between output and input). Another common parameter is the *group delay*, which is the derivative of the phase vs. frequency, $d\varphi/df$. In time domain this corresponds to the time delay between incoming and outgoing transition. While at first glance the group delay may seem unintuitive, it is easy to grasp when looking e.g. at a simple, ideal cable: Phase is a measure of the number of wavelengths that fit between input and output, and the wavelength in such an ideal cable is is inversely proportional to frequency – thus the group delay (and the time delay) is constant for all frequencies. Since a signal transition (edge) is a mixture of a wide band of frequencies, any deviation from this (meaning some frequency components will be delayed differently than others), as well as any variation in gain (loss) vs. frequency, results in a distortion of the signal edge.

behavior is called *dispersion,* and it results in signal distortion. All this makes it ill suited for digital applications.

- Chebyshev filter: Has some ripple (uneven loss vs. frequency behavior) in the pass band, but falls off much more steeply above its -3 dB bandwidth ("brick-wall behavior"). Just like the Butterworth filter *overshoot, ringing, and dispersion* make it a poor choice in the frequency domain.
- Bessel filter: Has already sizeable loss well below its -3 dB bandwidth, but on the other hand it has zero dispersion (it group delay is constant over frequency) and very little overshoot. It is frequently used in commercial time-domain filter designs.
- Gaussian filter: Very similar to the Bessel filter, it has minimum rise time with no overshoot or ringing, and the delay vs. frequency is largely constant.

An example as illustration: Let's assume we have a simple R-C filter with R = 50 Ω, C = 10 pF. This could be a signal from a coaxial cable driving a lower-performance oscilloscope probe. The time constant and rise time is then

$$T_{c,filter} = R \times C = 500\,\text{ps}, \quad T_{r,filter} = 2.2 \times T_{c,filter} = 1100\,\text{ps} = 1.1\,\text{ns},$$

respectively. The bandwidth of this setup is then

$$BW_{-3dB} = \frac{0.35}{1.1\,\text{ns}} \approx 320\,\text{MHz}.$$

1.5.3 Adding Rise Times

Let's say we now have two filters (rise times $T_{r,1}$ and $T_{r,2}$, respectively) in series. What will be the combined rise time, i.e. what will be the rise time of a very fast rising signal after it passes both filters? It turns out that for Gaussian filters (and in good approximation also for exponential filters and many other low-pass filters we are likely to have in our test setup) the rise times do not add up linearly, but as root-mean-square (RMS, geometric sum):

$$T_{r,tot} = \sqrt{T_{r,1}^2 + T_{r,2}^2} \qquad (9)$$

This is very fortunate, because the RMS sum increases slower than the linear sum, so our total bandwidth is higher. As a side remark, the same formula applies if we have (other than the assumption so far) a signal with some *finite* rise time $T_{r,signal}$ going through a filter – signal rise time and filter rise time add up according to above formula. This is easy to understand – we can always assume that the signal was originally rising infinitely fast and then went through another filter (with rise time $T_{r,signal}$) before it came to the filter under consideration. As an example, let's take the R-C filter from before (with a rise time of 1.1 ns) and send in a signal with a rise time of (a) 2 ns, (b) 100 ps. In the first case the output signal will have a rise time of

$$T_{r,tot} = \sqrt{(2\,\text{ns})^2 + (1.1\,\text{ns})^2} \approx 2.3\,\text{ns}, \tag{10}$$

which already shows some degradation due to the filter. But in the second case the rise time is

$$T_{r,tot} = \sqrt{(0.1\,\text{ns})^2 + (1.1\,\text{ns})^2} \approx 1.104\,\text{ns}, \tag{11}$$

in other words, the rise time is completely dominated by the filter. As a rule of thumb we can say that whenever one rise time is at least three times larger than the other, the other rise time is of no importance. (The opposite case to above example would be a filter that has a much faster rise time than the signal – the filter would then have no effect on the signal. In the frequency domain picture this means a low-pass filter with a bandwidth much higher than the signal bandwidth will pass the signal undistorted).

If we have a series of several filters in a row (with rise times $T_{r,1}$, $T_{r,2}$, $T_{r,3}$, and so on), the total rise time is calculated as follows:

$$T_{r,tot} = \sqrt{T_{r,1}^2 + T_{r,2}^2 + T_{r,3}^2 + \ldots} \tag{12}$$

Two notable and important exceptions to this formula exist (we will look at those effects in more detail shortly): Cables that show rise time degradation (in other words, limited bandwidth) due to skin effect or dielectric losses – and all cables do to some extent – do *not* follow the RMS rule. In other words, if for example we attach two cables of same type (rise time $T_{r,cable}$) and same length together, the resulting cable will *not* have a rise time of $T_{r,cable} \times \sqrt{2}$, but *more* (i.e. the bandwidth degradation will be worse). For dielectric losses the increase is approximately *linear* with cable length, while the skin-effect caused rise time increases *quadratically* with length.

1.5.4 Effects on Signal Propagation

Increased signal rise time is only one of several effects that low-pass filters have on our signals. Two others are reflections and delay. We will deal with reflections shortly when we will talk about parasitics in the transmission path. Why filters inevitably add delay (even in the case when they are geometrically negligibly small) is easily understood if we take a look at Figure 8 that shows the effect of a simple R-C low-pass filter: It is true that the output signal *starts* rising the very moment the input signal arrives, so we could be tempted to argue that there is no additional delay. But because of the rise time degradation caused by the filter its slew rate is lower. And in digital land "timing" is always the instant where the signal crosses a certain threshold (50% or midlevel in Figure 8), and the output signal – rising slower – takes more time to get there. This is meant when we talk about "filter delay".

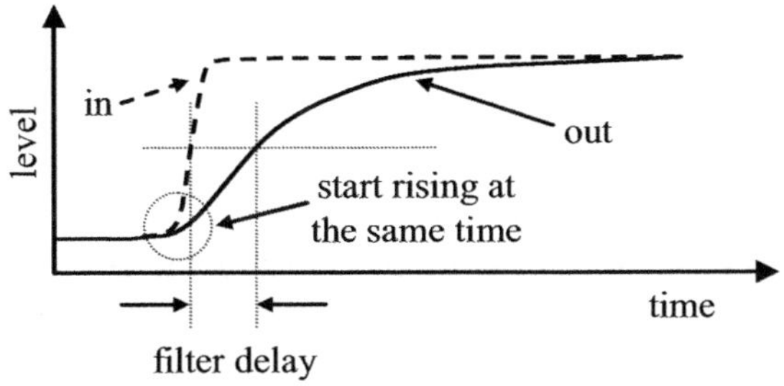

Figure 8: A low-pass filter (in this case a simple 1-pole R-C filter) distorts the signal and delays the time when the output signal crosses the threshold.

The exact amount of delay depends on a variety of factors – filter rise time, the exact shape of the filter response curve, signal rise time, and the exact edge shape. But for "well-behaved" filter and signal shapes (Gaussian and exponential shapes for example are "well-behaved") the main dependency is simply on the ratio of signal rise time to filter rise time. The delay here is always a fraction of the filter time constant $T_{c,filter}$ (which we remember to be approximately $T_{r,filter} / 2.2$):

$$delay = k \times T_{c,filter} = k \times \frac{T_{r,filter}}{2.2}, \quad k = 0.69 \ldots 1.0 \qquad (13)$$

Figure 9 displays the approximate dependency: For a very fast edge (much faster than the filter rise time) the delay is approximately 69% of the filter *time constant* (not the filter rise time), while for a very slowly rising edge it approaches the filter time constant. Though this increase (from 69%

to 100%) is usually small, it still means that whenever we have a filter in the path, the propagation delay becomes dependent on the signal parameters – there is no longer such a thing as *the* propagation delay!

Once again let's take our simple R-C filter from before (T_r = 1.1 ns) and send in the two signals with (a) 2 ns and (b) 100 ps rise time, respectively. From Figure 9 we determine the factor k to be (a) 0.69, (b) 0.87, and the delays to (a) 345 ps, (b) 435. This is already a difference of 90 ps, definitely not negligible in a high-accuracy measurement setup. It also makes clear that if we have to characterize or calibrate this setup, we should do this with signals that have a very similar rise time to the signals we want to measure!

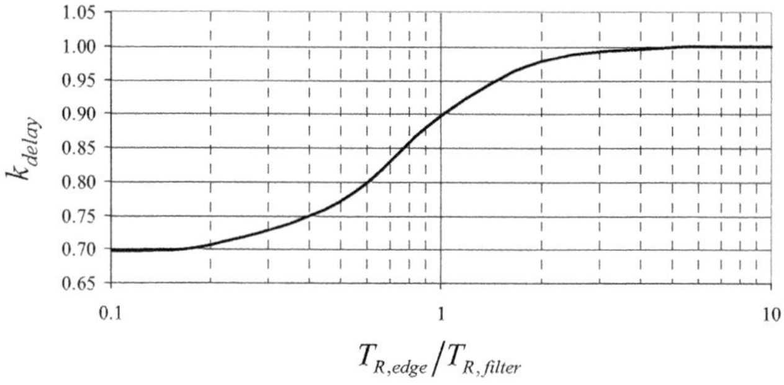

Figure 9: The delay of the output signal depends mainly on the ratio between signal rise time and filter rise time (or signal time constant and filter time constant).

1.6 Transmission Lines

One of the most important concepts in high-speed signaling is the transmission line. In plain English, a transmission line is a conductor arrangement where the capacitance as well as the inductance is well defined and (at least reasonably) homogeneous along the signal path. The easiest way to achieve this is to make the geometry along the conductors homogeneous. Coaxial cables are prime examples – the center conductor (carrying the signal current) is at a constant distance from the shield (carrying the signal return current) along the whole length of the cable, and both cross-sections do not vary over their length either; between center and shield is a dielectric material (i.e. an insulator) that is homogeneous as well. Traces on a PCB are another example. Two "free-flying" banana jack cables on the other hand would rarely be regarded as a transmission line – most likely their mutual distance is in no way constant along their length.

The importance of transmission lines lies in the fact that they are able to transmit very fast signals with minimum distortion and in a well-defined and rather easy-to-describe manner.

1.6.1 Key Parameters of Ideal Transmission Lines

Textbooks about transmission line theory usually start out with the easiest case – the perfectly homogeneous, loss-less transmission line – and we will do likewise.

We already mentioned that transmission lines are characterized by a homogeneous capacitance and inductance along their length. This is shown in Figure 10(a): Signal path and return path form an inductive loop with inductance L_{tot}, and together they also constitute a capacitance C_{tot}. So far, so good. But alas, another complication awaits us: Capacitance and inductance are not *lumped* (concentrated) at one point, but rather they are *distributed* over the length of the transmission line.

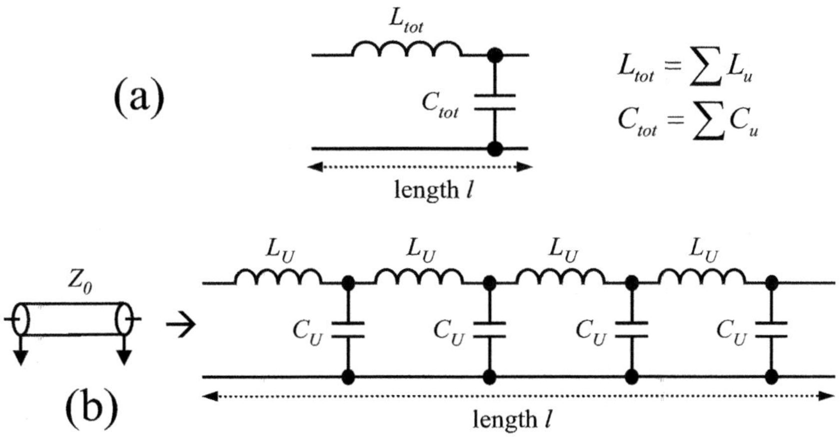

Figure 10: (a) Lumped model of a transmission line. (b) Discrete, distributed model of an ideal, loss-less transmission line.

Why is this important? As we know, the rise times (and even clock periods) of our high-speed signals are so fast that the signal has no way of traveling through the whole path during that time, and therefore there is no way it can "experience" the action of the total line capacitance and inductance. At any given point in time it only "sees" a small fraction of the line. Thus what becomes important is less the total line capacitance C_{tot} or inductance L_{tot} but rather their incremental values C_U and L_U *per unit length.* (As we will see shortly, the choice of what is the unit length – meters, inches, etc. – is actually irrelevant for our theory). C_{tot} and L_{tot} is what a low-frequency LCR-meter would measure[12] (as long as its frequency is so low

[12] To measure the C_{tot}, the line would have to be be open at the end. To measure the L_{tot}, the line would have to be shorted at the end.

that one period is many times longer than the propagation delay through the line. Hence the frequent misconception of looking at a cable as a "capacitive load" – sure, we can measure its total capacitance, but we will see shortly that this number is irrelevant for the propagation of the signal. We can model the distributed nature of the parameters by a long alternating chain of small capacitances and inductances, as shown in Figure 10(b): The sum of all capacitances and inductances equals C_{tot} and L_{tot}, respectively.[13]

The signal propagation speed v is always less (or at best equal) to the speed of light in vacuum (which is 3 x 10^8 m/s):

$$v = \frac{c}{\sqrt{\varepsilon_r}}, \tag{14}$$

where ε_r is the so-called relative dielectric constant[14] of the insulator material surrounding the conductors. Typical values for ε_r are between 2 and 5 for dielectrics used in cables and PCBs, and higher for ceramics or materials used in integrated semiconductor devices. Vacuum (and air for our purposes) has $\varepsilon_r = 1$.

To get an idea of typical values, let's consider a cable with $\varepsilon_r = 4$, carrying a signal of 1 GHz frequency (i.e. period 1 ns). During one period the signal covers a distance of

$$d = v \times t = \frac{c}{\sqrt{\varepsilon_r}} \times period = \frac{3\times10^8 \text{ m/s}}{\sqrt{4}} \times \left(1\times10^{-9} \text{ s}\right) = 0.15 \text{ m},$$

i.e. most likely less than our cable length. Even more extreme would be the distance covered during the rise time $T_r = 50$ ps of a 10 Gb/s signal: A measly 7.5 mm!

1.6.2 Reflections, Timing, and Signal Integrity

A surprising property of our ideal transmission line is that even though it consists entirely of capacitances and inductances (i.e. purely reactive components), it looks for an *incident signal* (i.e. a sudden voltage *change* at the input) like a perfectly ordinary ohmic resistance: There is no phase lag between applied voltage and the current that flows, and the ratio between

[13] In theory we'd have to use an infinite number of infinitely small elements, but for practical modeling as well as understanding it is sufficient to use a large enough number of small elements.

[14] Actually "constant" is a poorly chosen description because it varies with a host of parameters – temperature, humidity, signal frequency, you name it!

voltage *change* and current *change* is a constant – the so-called *characteristic impedance* Z_0. This impedance together with the *propagation delay* T_{pd} (the time it takes for the signal to make it all the way to the other end of the line) completely describes such an ideal transmission line:

$$Z_0 = \sqrt{\frac{L_u}{C_u}} = \sqrt{\frac{L_{tot}}{C_{tot}}}, \text{ and}$$

$$T_{pd} = \frac{c}{length \times \sqrt{\varepsilon_r}} = \sqrt{C_{tot} \times L_{tot}} = Z_0 \times C_{tot} = \frac{L_{tot}}{Z_0}, \tag{15}$$

$$(L_{tot} = L_u \times length, C_{tot} = C_u \times length).$$

As an example, when applying a 1 Volt *change* to one end of a line with $Z_0 = 50\ \Omega$ (the usual choice for the impedance of a digital transmission line), a current *change* of 20 mA will result. This signal wavefront will then propagate along the line with the propagation speed *v*. However, there are some important differences between a simple ohmic resistor and a transmission line: First, the signal will arrive at the other end of the line only after the delay T_{pd} (while a resistor does not have any delay). Second, no energy is dissipated, but it is stored in the electric and magnetic fields along the line. Third, at the end of the line a reflection of the signal will occur; more precisely, reflections occur whenever there is a change in the impedance along the line (and the end of the line can be seen as a change to infinite impedance). The relative portion of the signal that gets reflected and transmitted at a transition from the line impedance Z_0 to a load impedance of Z_L is given by the reflection coefficient ρ and the transmission coefficient τ:

$$\rho = \frac{V_{reflected}}{V_{incident}} = \frac{Z_L - Z_0}{Z_L + Z_0}, \text{ and } \tau = \frac{V_{transmitted}}{V_{incident}} = \frac{2 \times Z_L}{Z_L + Z_0} \tag{16}$$

Since the transmission line is loss-less, the total electrical energy is conserved:

$$\frac{V^2_{reflected}}{Z_0} + \frac{V^2_{transmitted}}{Z_L} = \frac{V^2_{incident}}{Z_0} \tag{17}$$

The load impedance does not have to be an ohmic resistor (i.e. real-valued and time-invariant), but can also be a capacitor, some network, or even another transmission line with different characteristic impedance. Three important cases can be derived from above equation: First, if the load impedance is equal to the line impedance ($Z_L = Z_0$), we speak of *matched termination*. The reflection coefficient disappears ($\rho = 0$), so no reflection occurs. This is the ideal case since it avoids any reflections that could interfere with our data signal. If the load impedance is infinite (meaning no termination at all), the coefficient is +1, i.e. the full signal is reflected back and will interfere with our transmitted data. The last case would be impedance equal to zero, i.e. a short to ground. The coefficient becomes −1, meaning again the full signal is reflected, but with opposite polarity. All other cases lie somewhere between those three extremes.

One thing we learn from this is the importance of keeping all our elements – cables, connectors, oscilloscope inputs, and so on – matched to each other in impedance (most likely 50 Ω in digital applications). The usual practice at low speeds to have the oscilloscope input at 1 MΩ impedance is a bad idea because it will cause huge reflections back into our system.[15] We should also be aware by now that referring to the scope probe (or cable to the scope) as a *capacitive* load – as done often for slower-speed scopes and probes – completely neglects things like propagation delay through the cable and reflection at its end (when the scope input does not have matched termination). We really have to regard every section of our signal path as a transmission line.

1.6.3 Parasitics in the Transmission Path

Localized deviations from a constant ratio between inductance and capacitance per unit length (which determines the line impedance) can be regarded as either excessive shunt capacitance or excessive series inductance at the particular location.[16] A good example is a connector where the design was not done very accurately, so the shield does not keep constant distance from the center conductor and as a result the whole assembly has too little capacitance (or too much inductance, for that matter). If the deviation extends only over a length (propagation time) smaller than a fraction of the rise time of the signal traveling down the line, then it is unnecessary to apply the full-fledged transmission line picture. Instead we can treat it as lumped at a point and we usually talk of inductive or capacitive "*parasitics*". Such parasitics are caused by any imperfection in the path geometry, like

[15] The only case were we do want high impedance is when we probe in the middle of an otherwise terminated line – here any additional load would cause an impedance mismatch on the line and thus reflections. But the probe has to be an active probe; as a simple coaxial cable to the scope – being longer than the signal rise time – would not work: What counts here is the cable impedance, not the termination on the scope side!

[16] Note that each section of a transmission line has of course a certain capacitance and inductance, but only the *excess* capacitance or inductance – the part that exceeds what is needed to obtain an impedance of Z_0, typically 50 Ω – will affect signal fidelity.

connectors, vias, narrow bends, etc., so minimizing those is one of the main tasks in good high-speed signal engineering.

Parasitics are unwanted guests since *they cause reflections, waveform distortion, and limit the path bandwidth.* The latter effect is easily understood if we recall that a transmission line acts as an ohmic load Z_0 to any transition, so if there is a capacitance C (or an inductance L) somewhere along the line, it forms an R-C or R-L filter with a time constant of

$$T_{R-C} = R \times C = \frac{Z_0}{2} \times C, \text{and } T_{R-L} = \frac{L}{R} = \frac{L}{2 \times Z_0}. \tag{18}$$

The factor ½ for the R-C filter comes about because the line leading to and the line leading from the capacitance act in parallel, giving a Thevenin equivalent source impedance of half the line impedance (Figure 11(a)). For the R-L filter on the other hand they act as two lines in series, effectively doubling the source impedance (Figure 11(b)). The 10/90 rise time $T_{10/90}$ and the 3 dB bandwidth BW_{-3dB} of such a filter is of course given by

$$T_{10/90} \approx 2.2 \times T_{R-C\,(\text{or}\,L-C)}, \quad BW_{-3dB} \approx \frac{0.35}{2.2 \times T_{R-C\,(\text{or}\,L-C)}}. \tag{19}$$

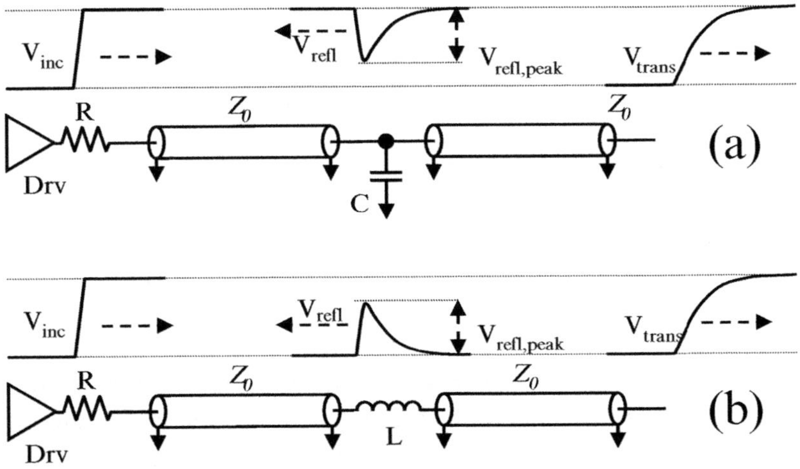

Figure 11: (a) Parasitic shunt capacitance in the path. (b) Parasitic series inductance in the path.

We already dealt with the detrimental effects of low-pass filters on our *transmitted* signal – increased rise time, distorted waveform, delay – in the previous sections, and low-pass filters caused by parasitics are no exception here.

Why parasitics cause reflections can be understood with the following simple consideration: If a transition hits the initially uncharged capacitance, the capacitance will "soak up" any charge it can get, effectively acting as a short to ground, i.e. with a reflection coefficient of -1. When it charges up more and more, the current into it decreases, until it is finally completely charged up and no longer has any effect on the signal (until the next transition arrives), i.e. a reflection coefficient of zero. In other words, there is a strong negative initial reflection that then decays over time. The exact height, shape and duration depend on both the size of the parasitic capacitance as well as the rise time of the incoming transition. Longer rise times smear out the response, so the reflection becomes shallower but longer. For parasitic series inductances it works very similar, except that it initially acts like an open (no current can pass, and we get a positive reflection spike) and then gradually opens up. Figure 11 also shows the reflected waveform for a parasitic capacitance (a dip) and a parasitic inductance (a peak), respectively.

Assuming the edge is a linear ramp there exists a closed analytic formula for the height of the reflected peak (or dip):

$$\frac{V_{reflected}}{V_{incident}} \approx \frac{T_C}{T_{rise}} \times \left\{1 - \exp\left(-\frac{T_{rise}}{T_C}\right)\right\}$$
$$\text{capacitive parasitics: } T_C = \frac{Z_0}{2} \times C \qquad (20)$$
$$\text{inductive parasitics: } T_C = \frac{L}{2 \times Z_0}$$

Note that the rise time here is $T_{0/100}$, which works out approximately to

$$T_{0/100} \approx 1.25 \times T_{10/90} \approx 1.67 \times T_{20/80}\,. \qquad (21)$$

As we will see later, a reflection of p percent means a timing error equal to approximately p percent of the signal rise time. Figure 12 shows a plot of the reflection amplitude versus the signal rise time according to above formula.

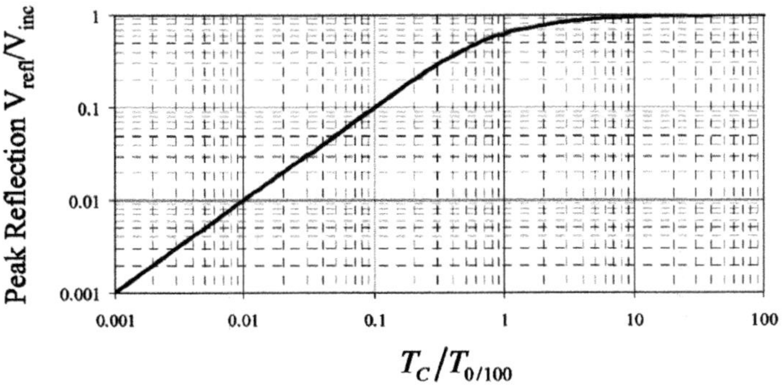

Figure 12: The size of the reflection caused by a parasitic element in the path depends on the ratio between the size of the parasitic (given by its time constant) and the rise time of the incident signal.

Let's say we want to probe the signal on a fast serial link, running at 3.2 Gb/s with a rise time of $T_{r,signal,10/90} = 100$ ps. Since we do not want to put an additional 50 Ω load on the middle of the line (this would guarantee to impact system operation negatively!), we are using a fast high-impedance FET probe with a specified input capacitance of $C_{probe} = 2$ pF. According to the formula above, attaching the probe will create a low-pass filter with a time constant $T_{c,filter}$ and a rise time $T_{r,filter,10/90}$ of

$$T_{c,filter} = \frac{Z_0}{2} \times C = \frac{50\,\Omega}{2} \times 2\text{ pF} = 50\text{ ps},$$

$$T_{r,filter,10/90} \approx 2.2 \times T_c = 110\text{ ps}.$$

In other words, just the fact of attaching the probe to the middle of our line will degrade the signal rise time to

$$T_{r,out,10/90} = \sqrt{T^2_{r,signal} + T^2_{r,filter}} = \sqrt{100^2 + 110^2}\text{ ps} \approx 150\text{ ps},$$

and in addition this will result in a sizeable reflection back to the driver of

$$\frac{V_{reflected}}{V_{incident}} \approx \frac{50\text{ ps}}{1.25 \times 100\text{ ps}} \times \left\{1 - \exp\left(-\frac{1.25 \times 100\text{ ps}}{50\text{ ps}}\right)\right\} \approx 37\%\ (!)$$

of the incident signal. Such a large reflection is almost bound to cause trouble, especially when the driver source termination is not exactly 50 Ω so the reflected signal gets re-reflected into our oscilloscope (and into the receiver).

1.6.4 Lumped vs. Distributed Elements

After dealing separately with transmission lines and then with parasitics we could ask ourselves: Now when does an element act as a transmission line, and when as a parasitic? Or, in other words, when do I have to go all the way and treat some structure (e.g. a short stub hanging off my path) as a transmission line, and when can I get away with treating it as a lumped inductance or capacitance?

This is indeed not a trivial question, and whole books are dedicated to answering it. The first remark should be that we can never go wrong treating something as a transmission line (though it may be a very inhomogeneous one), but it can entail unnecessary effort (unnecessary in the sense that the improvement in accuracy will be immeasurably small for the given purpose). On the other hand, treating something as a lumped element means (somewhat simplified) that we neglect its geometric size, or in other words the fact that a signal will take some time to traverse it from one end to the other. Clearly we can only do this if the propagation time is much smaller than the time scale of interest. A common rule of thumb, applicable in many, but not all cases, is that the propagation time must be smaller than 1/6 of the signal rise time – we can see that our model of the signal path will have to change when we want to transmit signals with different rise times. One limit of this rule of thumb is that it will not work for systems with very low loss (for this purpose termination can seen as very high loss, close to 100%, because it completely swallows the incoming signal) – e.g. a resonant circuit with high Q-factor may have small geometric proportions, but will ring for much longer than its one-way propagation time.

To get a feeling for the proportions, let's say we have a coaxial cable of 1 m length, with a dielectric constant of the isolator is $\varepsilon_r = 4$, that leads to an oscilloscope with high impedance input, and want to measure a 1 MHz clock signal with a rise time of $T_{10/90} = 200$ ns. The propagation delay through the cable is

$$T_{pd} = \frac{length \times \sqrt{\varepsilon_r}}{c_{light}} = \frac{1\,\mathrm{m} \times \sqrt{4}}{3 \times 10^8\ \mathrm{m/s}} \approx 6.67\ \mathrm{ns}.$$

The period was only given to confuse us, because as we remember it is really the rise time that determines the signal bandwidth BW:

$$BW \approx \frac{0.35}{T_r} = \frac{0.35}{200\,\text{ns}} = 1.75\,\text{MHz}.$$

In this case the rise time is thirty times larger than the propagation delay, so even though the oscilloscope end is unterminated the signal will bounce back and forth many times during the rise time and be settled long before the signal has risen to full level (since the device driver will provide at least some source side termination). In this case we can assume the cable to be a simple capacitance (at these speeds the typical drivers have rather large output impedance, so the capacitance is much more important than the inductance), and this is the reason while until a few years ago most people regarded scope probes as nothing more than capacitive loads). But already around 10 MHz this model begins to break down for a cable of this length, and we really have to look at it as a partially unterminated transmission line with reflections running back and forth.[17]

If on the other hand we are dealing with a fast 10 Gb/s serial bus with signal rise times of about 35 ps (i.e. a signal bandwidth of around 10 GHz), then anything longer than about

$$length = \frac{T_r \times c_{light}}{\sqrt{\varepsilon_r}} = \frac{35\times10^{-12}\ \text{s} \times 2\times10^{8}\ \text{m/s}}{\sqrt{4}} \approx 5\,\text{mm}$$

will act as a full-blown transmission line – i.e. even the short contact pin between our probe point and our active FET probe!

1.6.5 Lossy Transmission Lines

When a signal travels along a cable or other transmission path, the propagation is never perfect. The signal always experiences a certain amount of *loss*, meaning that part of the electromagnetic energy is dissipated into heat or radiated away from the signal path through some process or another. This manifests itself in a decrease in signal amplitude for periodic waves (e.g. our well-known sine waves), so we will e.g. measure wrong amplitudes if we have too long a cable between our signal source and our oscilloscope (or other test equipment). Unfortunately most losses increase with signal frequency, which causes distortion of the signal edges (as we will see in the next section) in addition to simple level attenuation.

Why are we so interested in losses – and parasitics, for that matter? As we will see in the next chapter, since both cause waveform distortions, they

[17] Of course the it will be advisable to set the scope input to matched 50 Ω termination to avoid reflections.

both result in timing errors and data dependent jitter – something to avoid when we want to make accurate timing measurements.

There are excellent books that deal entirely with signal propagation including losses and how to minimize them, and anybody who wants to learn more about high-bandwidth interface design can't do wrong reading those. We will go now through a brief overview outlining the major contributors, and things we can do to improve our measurement setup based on this knowledge.

As a general principle, whenever a loss mechanism is frequency dependent, it will affect (distort) the signal shape during a transition, introducing timing errors. This is easy to understand when we remember the connection between time and frequency – higher frequencies mean shorter times (after a transition) and faster change, respectively. So if the loss is stronger at higher frequencies, the beginning of a transition ("short time") is affected more than the end ("long time"), and the signal rise time will be increased.

1.6.5.1 Ohmic Resistance

Ohmic resistance is by far the most straightforward of all loss mechanisms. It is caused by the ohmic DC resistance of the conductors and it does not change with frequency.[18] Ohmic losses result in a constant reduction in signal amplitude, but fortunately – being frequency independent – they do not affect the rise times or overall shape of the transmitted signal.

Ohmic loss reduction is very straightforward – those losses increase linearly with trace (or cable) length, and are reduced by use of larger cross sections of the conductors, and by higher-conductivity material. For PCB traces, where copper is the material of choice most of the time anyway, which is an excellent conductor, possible improvements are limited; highest-purity material (electrolytic copper) can reduce the specific resistance somewhat. At the same time, making the traces very wide soon runs into routing space constraints on the PCB (unless additional layers are added, which increases cost, complexity, and forces longer vias with more parasitic capacitance), so the best option is to make traces as short as possible and use the highest feasible plating thickness. Similar considerations apply to coaxial or twisted-pair cables, we should always strive to make them as short as possible, and larger cross-sections (thicker cables) will improve things further.

[18] It may change with signal amplitude if the signal power is large enough to cause significant heating of the conductors, but it digital applications this is rarely the case.

1.6.5.2 Skin Effect and Proximity Effect

When the current changes over time (e.g. sinusoidal clock signal, or a digital signal transition), then the current distribution in the conductors is no longer homogeneous. Instead, mutual induction between all the partial currents across the cross section causes the current to be pushed out to the surface of the conductor. For frequencies above a few MHz virtually all current flows in a very thin layer below the surface, reducing the effective current-carrying cross-section and thus increasing the ohmic losses. For sine wave signals going through a cylindrical conductor the depth δ_{skin} of the current carrying layer decreases with the square root of the frequency[19]:

$$\delta_{skin} = \sqrt{\frac{\rho}{\pi \times f \times \mu_0 \times \mu_r}}, \tag{22}$$

where f is the frequency, ρ is the specific resistivity of the conductor, $\mu_0 = 4\pi \cdot 10^{-7}$ Vs/Am, and μ_r is the relative magnetic permeability of the conductor ($\mu_r \approx 1$ unless the conductor is ferromagnetic).

As a result the resistance (and loss) *increases with the square root of the frequency*. For conductor shapes and arrangements other than a cylindrical coaxial cable the trend with frequency is somewhat different (and in most cases no closed analytical solutions exist), but the $\sqrt{f}$ -trend is usually a very good approximation. In the time domain, time being inversely proportional to frequency, this means the losses disappear only with $1/\sqrt{time}$ (time measured from somewhere near beginning of the transition). This causes an increase in the signal rise time and – even more annoying – a long, very slowly vanishing voltage dropm, which makes skin effect one of the major causes for data dependent timing errors.

What is important to remember is that the rise time increase caused by skin effect is proportional to the *square* of the path length, i.e. all other things being equal a cable twice as long exhibits a rise time *four* times as large (or, in the frequency domain, it has only a *quarter* of the bandwidth). As a result, more than for any other loss effect keeping cable lengths to a minimum is absolutely vital for controlling this type of loss.

Since skin effect losses are nothing more than ohmic losses aggravated by a reduced effective cross section, some of the same considerations apply: Use the highest conductance material available, and make the traces and cables as short as possible. But there are differences as well: Because the current flows only on the conductor's surface, increasing the conductor

[19] Actually only $(1-1/e)$ of the signal flows within δ_{skin}, but the drop-off further into the conductor is exponentially fast.

thickness has only negligible effect – it is the surface that counts. Widening the trace (or, in the case of coaxial cables, increasing the cable's diameter) is one possible solution. What's more, the transversal trace dimensions (width, distance to ground plane, or for coaxial cables the distance between center conductor and shield) must be kept small compared to the signal's wavelength, otherwise "strange" non-TEM modes[20] will occur: Put into simpler terms, it means that the simple one-dimensional picture of the propagation breaks down, the propagation must be treated truly three-dimensionally, and the end result is dispersion and distorted edge shapes.

The proximity effect is related to the skin effect insofar as it is also caused by induction. But while for skin effect it is the self-induction between the partial current in the signal conductor, proximity effect results from the interaction between signal current and return current. Its effect is to make the current distribution inhomogeneous, with the larger amount of current flowing on the side closer to the other conductor, which also reduces the effective cross section and increases the ohmic resistance.

1.6.5.3 Dielectric Losses

The dielectric material, i.e. the insulator between signal conductor and return conductor (center conductor and shield in the case of a coaxial cable) consists of molecular dipoles. In a (very simplistic!) picture, those dipoles have to "turn around" whenever the polarity of the applied field changes. At the same time they experience some "friction" and so at each turn they lose some energy – which has to come from the electric signal energy – that gets converted into heat.[21] Even this simplistic picture predicts correctly that dielectric loss increases approximately linearly with frequency, so it, too, distorts signal edges.

The amount of dielectric loss per unit length in a given material at a specific frequency is a constant. So the only choices to reduce total signal loss are to shorten the path (a cure-all for any loss effect), and second, to use lower-loss materials.

Theoretically the best "material" would be of course vacuum, which has no losses at all; for signal integrity purposes, air is just as good. But airlines – used as impedance standards – do not make for very usable cables. All is not lost, though. High-performance cable manufacturers took up the idea that air is an almost loss-less medium: Instead of a solid layer of dielectric, they

[20] TEM (transversal electrical mode) denotes a situation where both electric and magnetic fields are at right angle to the propagation direction. This is the simplest form of propagation in a cable because the propagation is then one-dimensional along the cable's length (no complicated three-dimensional field distributions exist).

[21] The same effect heats food – acting as the dielectric – in a microwave oven!

produce a foamed dielectric (Teflon is a good choice) that consists to a large part of tiny air bubbles with a bubble size much smaller than the signal's wavelength. This reduces the effective (average) loss factor.

Just as skin effect, dielectric loss increases the signal rise time. But now the increase with cable length is linear (i.e. doubling the cable length doubles the rise time and cuts the bandwidth in two) and thus less dramatic than for skin effect, but still significant.

1.6.5.4 Radiation and Induction Losses

Any conductor carrying high-frequency signals inevitable acts as an antenna that emits part of the signal into free space. A second loss mechanism is induction into nearby conductors, which also carries away energy from the signal. Again those effects are frequency dependent, i.e. they will degrade the signal shape. But while very important for EMI (electromagnetic interference) compliance, radiation and induction losses only have negligible effect on the signal integrity in usual cables and transmission paths, even for frequencies as high as 10 GHz.

1.6.6 Effects of Parasitics and Losses

Figure 13 displays a schematic view of how the total loss of a transmission path changes with frequency.[22] At designs for a few 100 MHz, skin effect losses are still dominant, but because the skin effect only increases with the square root of the frequency, while dielectric loss goes linearly with the frequency, above some frequency the latter loss mechanism becomes dominant. However, the transition region is very broad, so in today's designs we have always a mixture of both. PCB traces tend to have substantial dielectric loss, while the loss in high-performance cables is usually dominated by skin effect.

When a signal transition (an "edge") travels down the transmission path, losses, parasitics and reflections modify and degrade its shape, each type in a different way: Ohmic DC loss merely reduces the signal amplitude by a constant factor, but leaves the edge shape intact. Dielectric loss as well as parasitics increase the signal rise time, but disappear quickly (exponentially) after a transition. Skin effect on the other hand not only increases the rise time, but also it settles so slowly with time that it causes an additional voltage drop for a very long time after the transition. Parasitics form low-pass filters, again increasing the rise time, but their rise times add as root-

[22] The frequency scale should not be taken too literally as the exact ranges will depend on design and dimensions of the path; it is only meant to give a general idea.

mean-square. In addition, they cause reflections, which can interfere with subsequent transitions.

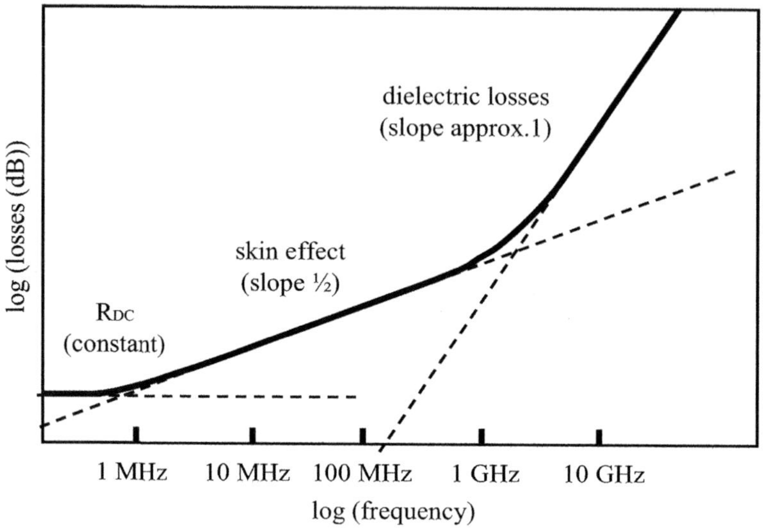

Figure 13: Transmission losses – both skin effect and dielectric loss – increase with frequency. Above a certain frequency dielectric loss becomes dominant.

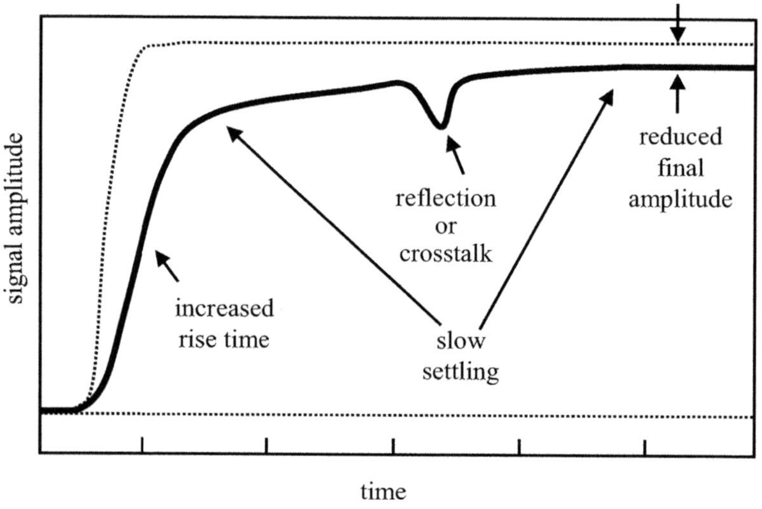

Figure 14: A signal transition degraded by ohmic losses, skin effect, dielectric loss, and parasitics, compared to the original (incident) signal shape.

Since rise time and bandwidth are inversely proportional, one can always view a rise time increase as a reduction in effective path bandwidth. Figure 14 shows a transition on the output of an imperfect, lossy transmission path, degraded by reflections and different loss types, compared to the clean input signal sent into the path.

1.6.7 Differential Transmission Lines

In recent years differential signaling has made large inroads in high-speed transmission schemes. Estimates are that a few years from now almost 100% of all PCB's will have at least some differential signal paths on them. There are several reasons for this: On one hand keeping signal and return paths symmetrical and close together greatly reduces electromagnetic emissions because their far fields cancel out. Second, since in such a transmission scheme the total signal current over the two lines of a differential pair is constant (at least as long as no differential skew is present), the current spikes drawn by the drivers are reduced by an order of magnitude, reducing power supply noise and ground bounce. Third, differential transmission is much less sensitive to residual ground bounce or external influences than single-ended signaling because the influences on the two lines of a pair largely cancel out.

For test and measurement purposes this means we have to concern ourselves with that topic as well, but the emphasis here is on accurate measurement results rather than optimizing the performance in an end application.

In contrast to wide-spread belief, there is nothing really special required *per se* for two lines to be "differential" – the only distinctive feature is that the signals the two lines carry are not independent, but are always complementary to each other (when one line is high, the other one is low, and when one line transitions, the other transitions in the opposite direction). Things like "coupling" and "differential impedance" (see below) are the result of specific design techniques associated with differential signaling rather than prerequisites for it.

When two transmission lines come very close to each other, their electric and magnetic fields start to overlap and induce voltages and currents into each other, interfering with the original signals on the lines. In "normal" (single-ended) signaling this referred to as capacitive and inductive crosstalk, and it is usually an unwanted feature. Another way to look at it is that the two lines have some mutual capacitance as well as mutual inductance which, depending if there is a transition on the other line, adds to or subtracts from the self inductance and from the capacitance against the ground.

However, if those two lines form a differential pair, then the transitions on them are no longer independent, and the crosstalk has always the same effect for each transition. In this case we talk of "coupling", but it is really just the same physical phenomenon. Which way the signals are influenced depends on the relative polarity of the transitions: If they are of opposite polarity ("odd mode"), the effective capacitance is increased, the effective inductance reduced, and thus the effective "odd mode impedance" Z_{odd} is reduced as well. For same-polarity transitions ("even mode"), the situation is exactly the other way around, causing the "even mode impedance" Z_{even} to be higher than the impedance of an isolated line. Those two impedances are related to differential impedance Z_{diff} (the total impedance seen by the differential signal, for which the two lines are effectively in series) and common impedance Z_{common} (where the two lines act in parallel) by simple formulas:

$$Z_{diff} = 2 \times Z_{odd}, \quad Z_{common} = \frac{Z_{even}}{2}, \quad Z_{even} \le Z_0 = \sqrt{Z_{odd} \times Z_{even}} \le Z_{even}. \quad (23)$$

The equality between Z_{even}, Z_{odd} and Z_0 happens when the lines are completely uncoupled. (A good example is the case when the two signals run in two independent coaxial cables – in this case we can forget about all those impedances and simply use the single-ended impedance – the differential impedance is then given by the two cables in series, i.e. 100 Ω when the cables are standard 50 Ω cables.) The beauty of this concept is that the transmission equations for differential and common signals stay the same as for single-ended signals as long as one replaces the impedance with the proper value for each case.[23]

In real-world applications close line spacing (resulting in considerable coupling) is used out of several reasons: First, it reduces emitted radiation – very important for a device in order to be compliant to EMI rules and regulations. Second, it makes differential lines less susceptible to external fields because those influences will cause the same disturbance in both lines of the pair, so they cancel out in the differential receiver. Third, routing both lines close together automatically means their propagation times will be well matched, so the differential signal integrity at the receiver is conserved. Fourth, since crosstalk (coupling) between the two lines is of no concern, routing the lines close together conserves board space, allowing either for smaller boards or less layers, thus decreasing board cost and/or size. In summary, we see that close coupling is merely a side effect of other

[23] To be precise, this simplification is only possible for *symmetric* differential pairs, but this is what one encounters 99% of the time anyway.

considerations in connection with differential signaling, but otherwise not an inherent necessity.

On the other hand, during test it can create a host of problems because it makes the impedances for common, differential, and single-ended signals different from each other. If coupling is different for some sections of the path (e.g. because the first part consists of uncoupled coaxial cables within the tester, and the second part is routed through coupled traces on a PCB), a path designed for differential signals will have impedance mismatches for single-ended signals. This can wreak havoc with test accuracy if – what is often the case – path delay (deskew) calibration is done with single-ended signals. Moreover, minimizing electromagnetic emissions or assuring minimal susceptibility normally takes a back seat in testing compared to maximum accuracy and clean signaling. Therefore a better practice *for test purposes* is to route signals sufficiently far apart, so there is no coupling and thus no dependency of the impedance on the exact trace separation. As to terminating a differential line, we'll see how this is done in the following section.

1.7 Termination

We already know by now that good impedance control throughout the path is extremely important to avoid reflections and maintain good signal integrity. But while mismatches along the line are usually in the range of just a few percent, the worst offender is the end of the path where the receiver (e.g. oscilloscope, or tester comparator) sits. By itself, receivers (or comparators, if it is the tester end) have high impedances, so the end of the line would be virtually unterminated (resulting in 100% reflection) if there weren't some additional termination. In a practical setup there are several schemes to terminate the signal at the end of the line.

1.7.1 Diode Clamps

In slower test equipment (below 100 MHz) and in slower-speed transmission schemes (e.g. SCSI), reflections and ringing on the transmission path was mitigated through the use of diode clamps (see Figure 15). In an idealized picture those clamps would clip any overshoot that exceeds one diode drop above (or below) the clamp voltage. But real diodes open up rather gradually with increasing voltage, they have only finite switching time, and together with the diode drop this leaves residual reflections. Another way to look at this is that diodes are nonlinear elements, i.e. their impedance changes with different biasing, so we cannot hope to maintain matched termination at all times – there will always be considerable residual reflections, so at best we can hope to *mitigate* the reflections, not to avoid them. All put together, diode clamps cease to be an effective termination method when speeds exceed a few 10 MHz. They do have their place in practical, lower speed bus applications, but are not recommended for a *measurement* setup where precision is more important than cost.

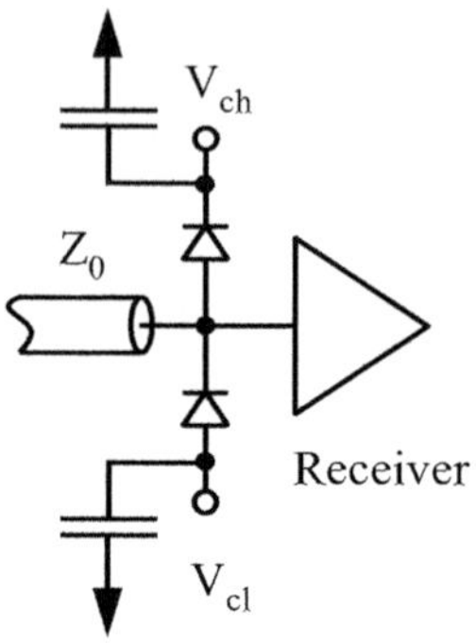

Figure 15: Diode clamp termination.

1.7.2 Current Loads (I-Loads)

A more sophisticated termination scheme is the so-called active load, also called current load or I-load (see Figure 16), and it is very common in automated large-scale production testers, but one is unlikely to encounter it in benchtop test equipment like oscilloscopes or bit error rate testers. It consists of a 50 Ω termination resistor connected to a diode bridge, which is supplied by a current source and a current sink and referenced to the termination voltage.

As long as the programmed source or sink current is not exceeded, the diode drops along the bridge keep the back end of the resistor at the same potential as the termination voltage. Thus in this range the load acts just like a 50 Ω resistor connected directly to the termination voltage, and will provide very effective matched termination. The behavior changes when the maximum current is reached – the effective termination impedance is no longer kept constant, and residual reflections occur. If the current limit is set to very small values, the I-load acts almost as an open, for large values (preventing it from ever reaching its current limits for the applied signal) it acts just like a matched 50 Ω termination all the time, suppressing all reflections, and for current limits between those two extremes it provides partial termination. Since for high-speed testing *all* reflections should be avoided, the "matched 50 Ω" mode of operation is usually the only one of interest, so the additional complexity compared to a simple 50 Ω resistor to the termination voltage is of little use here.

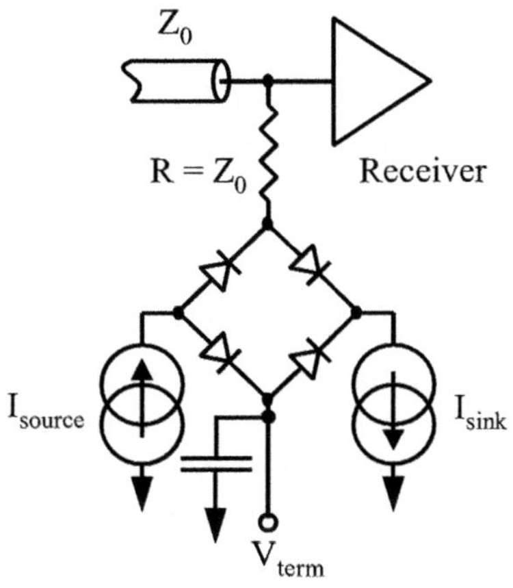

Figure 16: Termination with current loads.

1.7.3 Matched Termination (Resistive Load)

The function principle is almost trivial, as it consists of just a 50 Ω resistor connected to a DC voltage source, which – as we have seen previously – removes reflections completely, at least in theory. If available, this should be the termination of choice for high-speed testing. The challenge is to make the reaction time of the source smaller than the signal rise time, which is achieved through capacitive decoupling, as indicated in Figure 17.

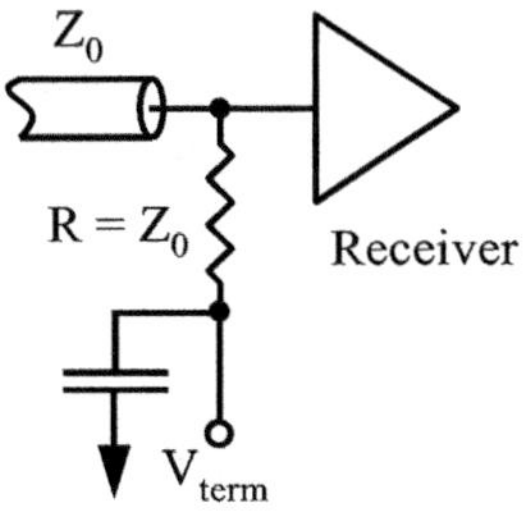

Figure 17: Termination with a matched resistive load. The voltage source needs to have fast reaction time to be effective (indicated by the capacitive decoupling).

A well-designed resistive load will leave virtually no residual reflections even at data rates in the Gb/s range. Fortunately virtually all high-speed test equipment like oscilloscopes, time interval analyzers or bit error rate testers already offer inputs terminated to 50 Ω as a standard, but usually hardwired to ground. Only slow-speed test equipment can get away with unterminated inputs. In case we need to terminate to some other voltage instead of ground, things become a bit tricky – we'll see later how to solve this.

A small but important detail – especially when we have a setup where we need to provide our own termination because the receiver is internally unterminated – is where to place the termination resistor. Figure 18 shows two different layout possibilities. Both seem to be valid solutions at first. But the first version (Figure 18(a)) has a stub between termination resistor and receiver, and the receiver is unterminated, so we will have strong ringing at the stub. Thus this is only an option if the propagation time through the stub is much smaller than the signal rise time, an unlikely case for high-speed applications. On the other hand the second solution (Figure 18(b)) does not have this shortcoming. Its only imperfection is the parasitic capacitance of the receiver, which is there in any case, and moreover its influence is reduced because the effective (Thevenin) source impedance is cut by half because of the termination, reducing the parasitic time constant by the same amount.

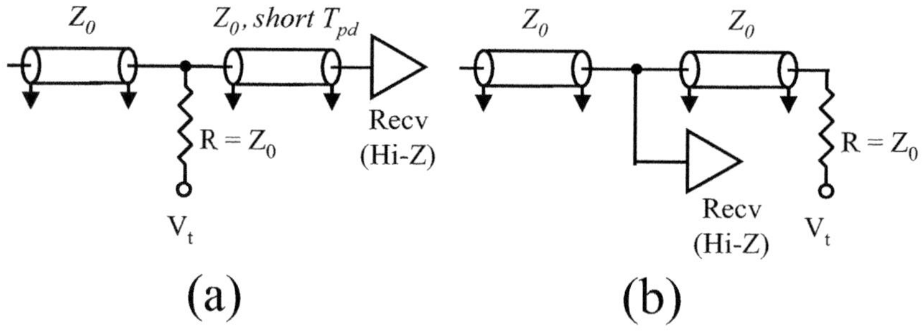

Figure 18: Different layout possibilities for resistive termination: (a) Residual stub will cause ringing (reflections), but layout takes less space. (b) Optimum signal integrity, but increased routing space.

1.7.4 Differential Termination

So far we have only talked about termination of single-ended signals. Differential signals add another layer of complexity to our considerations, because now we may have *two, unequal* characteristic impedances – one ($Z_{diff} = 2 \times Z_{odd}$) for the differential component and one ($Z_{common} = Z_{even}/2$)

for the common mode, so in this general case a single transistor will not suffice.[24]

For the general case (Z_{diff} and Z_{common} *are* different, for example for closely coupled lines on a PCB) several termination schemes are in use, shown in Figure 19, all based on resistive termination:

The two most elaborate schemes (Figure 19(a) and (b)) use both a set of three resistors, and they can match both differential and common mode impedance. Note that if the lines are uncoupled (and thus Z_{even} is equal to Z_{odd}), the bridging resistor in Figure 19(a) becomes infinite and we are back two single-ended lines with matched termination, while Figure 19(b) the center tap resistor becomes zero, which again makes the scheme identical to single-ended termination. As far as test applications are concerned, these two termination schemes can successfully remove reflections of the differential as well as of the common mode component of the signal. Their disadvantage lies in the necessary number of components (the total footprint requirements can create trouble). In addition, such a termination (if inside the tester) prevents any single-ended usage of the two channels, so this is something mostly used in benchtop applications where the emphasis in on maximum accuracy and signal integrity, but less in large scale automated test equipment.

A single-ended scheme (Figure 19(c)) can only perfectly terminate either the differential (the usual choice) *or* the common signal. If the termination resistance matches Z_{odd}, then the common signal will only be partially (but to a large extent) terminated, unless the lines are uncoupled so that Z_{even} and Z_{odd} are equal. A big advantage – not offered by any other differential termination scheme presented here – is that channels terminated this way can be used either in singled ended or in differential mode, greatly improving the flexibility of an automated test platform.

On the other end of the complexity spectrum is the simple bridged termination (Figure 19(d)), which consists of just a single resistor between the two lines, matched to the differential impedance. It provides full termination for the differential signal component, but none whatsoever for the common component. Due to its simplicity and small footprint this is often the method of choice for on-chip termination, but only to a lesser extent for the termination inside test equipment, where space (and cost) constraints are not as pressing compared to accuracy and signal integrity

[24] Fortunately, as we mentioned before, if the two signal lines are uncoupled, e.g. because they are running in separate coaxial cables or are otherwise isolated from each other sufficiently to avoid any crosstalk between the two lines, even and odd mode impedance will be both equal to the single-ended impedance, and as a result single-ended termination is all that is needed.

requirements, unless the goal is to match the final application with all its shortcomings.

One can improve the last setup by splitting the termination resistor in two and adding a decoupling capacitor in the middle (Figure 19(e)): The capacitor acts as an AC ground that can terminate spikes of common mode (and spikes are what we care most about, since static common mode offsets do not add data dependent timing errors). This is essentially the same as the elaborate scheme shown in Figure 19(b), but it omits the center tap resistor (which is very small anyway) and avoids its static power consumption (which can impact the common mode bias of the driver and/or the receiver).

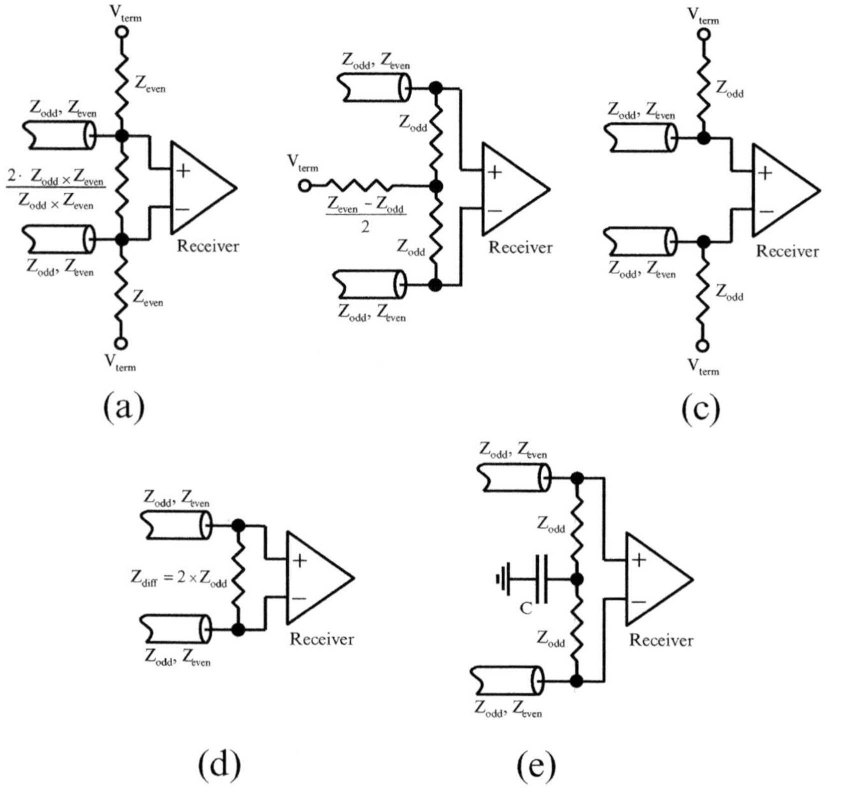

Figure 19: Termination schemes for differential lines: (a) ideal π-type, (b) ideal T-type, (c) single-ended, (d) bridged, (e) bridged with AC coupled center tap.

Chapter #2

MEASUREMENT HARDWARE

2. OSCILLOSCOPES & CO.

2.1 A Short Look at Analog Oscilloscopes

Most engineers are already familiar with the functions and the operations of traditional analog oscilloscopes, and since those legacy instruments have very limited use in high-speed timing and jitter measurement, we will only have a very cursory look at them, and also cover only those properties and functions that have applications in those types of measurements.

Figure 20 shows a high-level schematic view of a typical analog scope. The input signal is fed into an amplifier (or, for high signal amplitudes, into an attenuator). The normalized signal is split up, with one part going directly into the display system where it is applied to one pair of electrodes of the cathode ray tube to deflect an electron beam vertically, and a second part goes into the trigger system. Of course it is possible that such an oscilloscope has more than just a single channel (two to four is common); in this case the user can select which channel shall provide the trigger.

The trigger system is a slope-sensitive comparator (slope sensitive meaning it can either react to rising edges only, or to falling edges only) with adjustable threshold level. Whenever it detects a trigger edge (i.e. an edge of the appropriate polarity that crosses the threshold), it starts a ramp generator that sweeps the electron beam over the screen horizontally. After the sweep the generator goes back to its initial state and waits for the next signal from the trigger circuit.

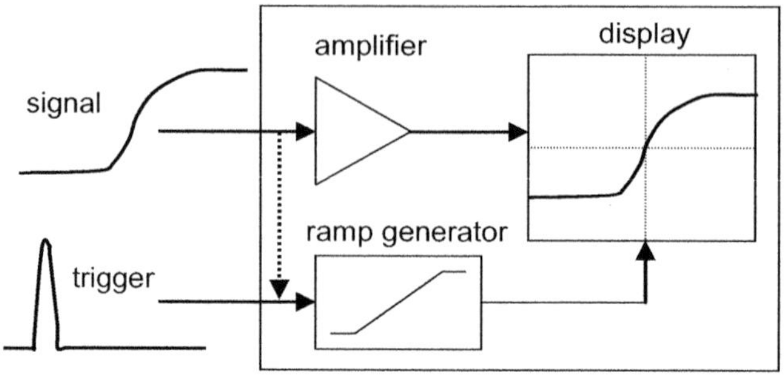

Figure 20: High-level block diagram of a traditional analog oscilloscope.

While the ramp generator sweeps the beam over the screen horizontally at a constant speed, the amplified input signal deflects it vertically in proportion to the signal amplitude at any given instant. Signal variations in time are therefore translated into different vertical positions along the curve. Wherever the electron beam hits the screen, the phosphor coating inside the screen glows for a short time, so the trace can be observed by the user. Normally the glow disappears after a very short time, so in order to obtain a steady image the signal has to be repetitive with a sufficient repetition rate (several Hz at least). The more often the beam hits a certain position on the screen, the brighter it glows. It is fair to say that in an analog scope the storage medium for the measurement is the electron tube's screen, although it is a highly volatile one.

The applicability of such an analog scope is limited by several factors: First, the deflection electrodes in the cathode ray tube have considerable capacity against each other, so do display very high frequencies – where those electrodes have to be charged and discharged very rapidly to follow the signal – very strong and very low impedance drivers are needed. Due to this effect it is very difficult to build an analog scope with a bandwidth exceeding 1 GHz.

Second, the signal information is solely stored in the glowing image on the screen (and that for a very short time), and cannot easily be translated into numbers and processed in a computer, e.g. to apply averaging, or to extract statistics on the edge positions for jitter analysis.

Third, it is difficult to capture very rare events because the screen will be dark most of the time. (One way around is to take a photograph of the screen with very long exposure time, but this requires minimum background glow, and the cycle of exposure – development – analysis is very slow).

Fourth, since the trigger event *starts* the sweep, we cannot observe what happened *before* the trigger – which may be important because it could tell us what led to some peculiar event. A way out of this is to use a passive analog delay line (a fancy word for a long, low-loss cable) to delay the signal so the trigger comes with a sufficient head start before the signal arrives, but losses in the cable prevent us from making this delay very large (more than a few 10 ns). At the same time it is equally difficult to look at things a long time after the trigger, because the whole interval has to fit on the screen and so large time delays mean our timing resolution will suffer.

For all those reasons analog oscilloscopes have all but disappeared from applications geared towards high-speed, high-accuracy measurements, and today are mostly used in low-range troubleshooting situations where their comparably low price and simple and intuitive operation outweigh their restrictions. One advantage they hold is that as long as the trigger repetition rate is high enough, no high-performance hardware is needed to provide fast screen refresh rates (and so see even elusive glitches), while until recently especially low-end digital scopes often had rather slow screen refresh rates due to poor computing power, so many "old-timers" among the engineers prefer analog scopes since they feel these "really show the signal as it comes".

2.2 Digital Real-Time Sampling Oscilloscopes

With the advent of microprocessors, integrated semiconductor memories, and fast analog-to-digital converters starting in the early seventies, a new type of oscilloscope has over time become the mainstay in most engineering: the digital storage oscilloscope.

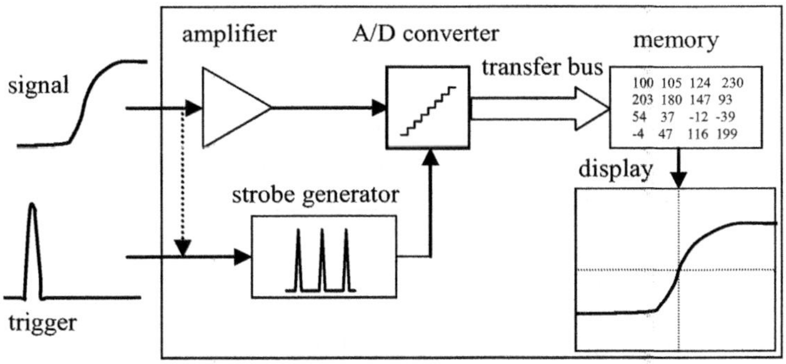

Figure 21: High-level block diagram of a digital sampling oscilloscope.

Looking at the general block diagram of such a scope (Figure 21) we can discern many similarities with an analog scope, but also a number of differences: Just as before, the incoming signal passes through an amplifier/ attenuator that makes its amplitude suitable for the subsequent stages. And again some part of the signal is fed to the trigger circuitry. But here is where the similarities end, at least what concerns the internals.[25]

The first big difference is that the signal is not fed directly to a cathode ray tube but instead goes into an analog-to-digital converter (ADC). This component *samples* the signal at regular intervals and so translates the incoming (analog) voltage into a stream of binary (digital) numbers that get then stored in a fast acquisition memory. From there the numbers are read by the scope's microprocessor and – possibly after a lot of mathematical manipulation like averaging, scaling or more advanced operations – displayed as a waveform on the screen. Thus the screen no longer has the purpose of information storage – the display is more like a side effect rather than a crucial part of the scope's operation. (If we wanted, we could even read the digital waveform information into an external computer and process or store it there without ever displaying it).

Second, since the data is available in digital format (voltage vs. time in regularly spaced intervals), it is no problem to apply whatever complex mathematical processing to it – averaging, interpolation, edge searches, measurements of frequency, amplitude, rise times and so on, some of which we will discuss later in this chapter.

The humble trigger has also experienced a big jump in complexity. While in the analog scope all it did was to *start* the sweep (acquisition), it now more or less directly *stops* it (or, we could also say, it acts like a filter that decides which sampled data to keep and which to discard), as will become clear with the description below[26]. Figure 22 illustrates the process graphically:

1. The acquisition (sampling) process itself runs without interruption, and data is transferred continuously into the capture memory. When the end of the memory is reached, the storage goes back to the beginning, overwriting the oldest captured data, so we can see the capture memory as a cyclical buffer of a certain length (a few hundred to several million samples is typical).
2. When the trigger event is detected, the acquisition process continues to run for a certain time (see below) and then stops (this is what we meant with "the trigger stops the acquisition"). The captured data – also called a "record" – is transferred to the scope's main memory and the trace displayed on the screen.

[25] The manufacturers of digital oscilloscopes usually make every effort to hide those complexities and make it look and perform as close to an analog scope as possible as far as the basic handling is concerned.

[26] We should note that the details may vary depending on scope manufacturer and scope model, but the outline below gives a good idea of how it is done in principle.

3. The system is now ready to capture the next trace and continues with 1.

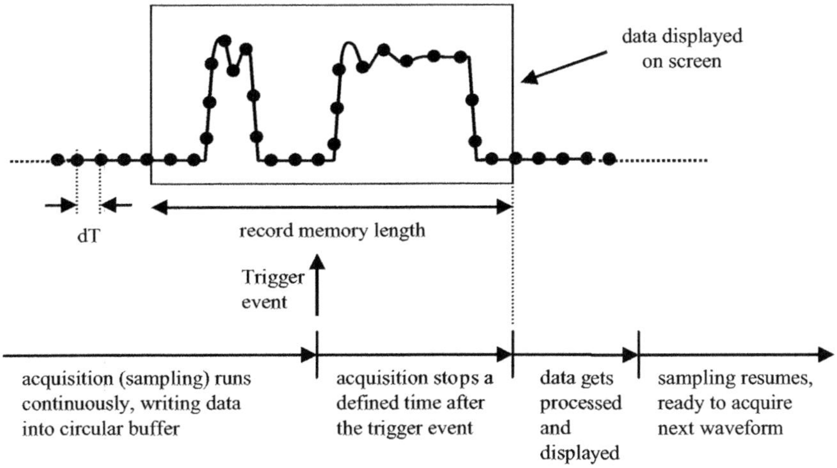

Figure 22: Principle of operation for real-time sampling.

We can select how long the system continues to captures after the trigger by appropriately setting the *trigger position*.[27] For example, if we set the trigger to the very beginning of the record, then the operation is identical to an analog scope – for the user the trigger seems to start the sweep (even though internally the acquisition has already been running before). This setting is also called "100% post-trigger". Usually this is the earliest position that the scope's software allows, even though in principle nothing prevents it from doing e.g. 200% post trigger, i.e. waiting even longer before it stops the sampling (in this case we would not see the trigger time on the waveform).

The other extreme would be to set the trigger position to the end of the record (i.e. "100% pre-trigger"), meaning the scope stops acquisition immediately when the trigger impulse arrives. In this case we see the waveform that happened before and all the way up to the trigger. This is something an analog scope cannot do. Finally, of course every setting between the two extremes is possible, e.g. 50% pre-trigger where the trigger position is in the middle of the record and we see some part of the waveform before as well as some part after the trigger point.

One weakness of this acquisition process is the time required to transfer the data record from the capture memory into the main memory, to process and display it. During this time no data is captured, and this time span may be many times longer than the time required to capture the record. In other words, we may only get a few snapshots of the waveform per second, which

[27] To keep things more comparable to analog scopes, this setting is usually called "horizontal position".

is not always obvious because the screen seems to get updated constantly (but our eyes cannot distinguish between 100 waveforms per second and 10000). If there are rare events (e.g. a glitch) in the signal, chances are high we will never see it at all, while an analog scope would show every single repeat and thus probably produce a faint but visible trace even of this rare event. Some of the highest-end oscilloscope designs seek to improve this situation by having dedicated data transfer and display circuitry (or don't update the screen at all during capture), so the main processor's performance does not get bogged down by those routine tasks. Some are able to capture hundreds of thousands of waveforms per second that way, which is close but not identical to a real analog scope in that respect. Even better, high-end digital scopes have very powerful trigger capabilities that allow triggering on specific features of the signal(s):

A big advantage of real-time scopes[28] as compared to analog scopes is their ability to capture single-time events. For that they also implement very sophisticated triggering schemes – on a good scope we can trigger on pulses that are lower than normal ("runt pulses"), too short ("glitches") or too long, are preceded by a specific bit pattern ("signature triggering") and so on. While important and extremely helpful for troubleshooting digital circuits, those modes are of less importance for standard timing and jitter measurements where the signal itself is usually repetitive, well defined and stable and we aim to characterize its properties with utmost accuracy.

A challenge for any real-time scope is to achieve the highest sampling rate (measured in GSamples/s) possible, so it can accurately capture even fast changing signal (more details to that later in this chapter). Today's leading-edge instruments achieve up to 40 GSamples/s, which by itself is quite an impressive feat, but even that means the interval between samples is 25 ps or longer, a lot of time with high data rates (keep in mind that 10 Gb/s means bit periods of only 100 ps, so even those fast – and expensive – scopes get only four samples per bit). It seems that real-time scopes have a hard time keeping up with those ever-increasing signaling speeds, especially now with the proliferation of ultra-high-data-rate serial transmission schemes.

Since the ADC has to acquire every sample in a very short time, there is not much leeway for it to settle out and obtain a highly accurate reading. Thus the resolution is usually limited to 8 bit (256 values, or a maximum resolution of just over 0.4%), and on some older models as few as 6 bit (64 steps). Even so the fastest scopes need to interleave two or more

[28] Referring to the previous paragraph, it seems that the "real-time" part of "real-time sampling scope" today has come to mean that it can capture a single waveform in a single shot, not that it necessarily processes or displays waveforms in real time as they come (which was the original meaning).

samplers (ADCs) to obtain the highest sample rates. Because of the limited vertical resolution it is important to make best use of it by adjusting the amplifier/attenuator so the signal covers close to full vertical (voltage) range of the sampler. Typically (on almost all scope models) this is the case when the displayed signal fills out the full vertical span on the display. But don't attempt to go further and have the signal exceed this range, no matter how little – this will most likely overdrive the input amplifier so it goes into saturation, and all bets are off with regard to waveform fidelity in this case! (That said, many scopes have designed in a small dynamic reserve of a few percent, but not more).

At the same time, to be able to capture long data streams and look at slow-changing effects even in fast data streams (that need high sample rates), the scope also needs a very large capture memory. The best ones available today thus come with many Megabytes worth of this memory, while lower-end scopes may max out at a few 1000 samples.

2.3 Digital Equivalent-Time Sampling Oscilloscopes

When dealing with very high frequencies (either fast data rates, or fast rise times, or both), normal real-time sampling scopes very soon hit two important barriers: First, they cannot increase their sampling rate beyond a certain limit, because the necessary data transfer rate into the capture memory gets unrealistically large, and even more important the ADC does not have enough time to settle between samples; speed and sampling accuracy/resolution are mutual tradeoffs. Second, as any other amplifier also the scope's input amplifier has some limited bandwidth – the best in class achieve around 12 GHz, i.e. enough to measure accurately signals up to maybe 3 GHz (or 6 Gb/s data rate assuming double-data-rate signaling).

So-called equivalent-time sampling scopes can often be employed for applications where the sampling rate and/or bandwidth of real-time scopes are insufficient. A fundamental difference is that the highest-performance versions do away with the amplifier, so the signal directly hits the sampler (if we ignore the termination resistor that is always present in those scopes to provide good signal integrity). This arrangement gets around the bandwidth restrictions of the amplifier for the price of largely reduced dynamic range because there is no way to scale the voltage range of the sampler. Analog bandwidths exceeding 70 GHz are possible[29], while the typical dynamic range is just around 1 V (normally this can be on top of some selectable offset of maybe +/-1 V).

[29] The highest-bandwidth commercially available sampling head that the author is aware of exceeds 100 GHz.

Of course physical limitations for the sampling rate still hold true, so those oscilloscopes do not even attempt to sample the signal in one sweep. Instead they rely on the assumption that the signal to be measured is repetitive, and they acquire only one sample per repeat and put them together to reconstruct the original waveform.[30] This process is illustrated in Figure 23.

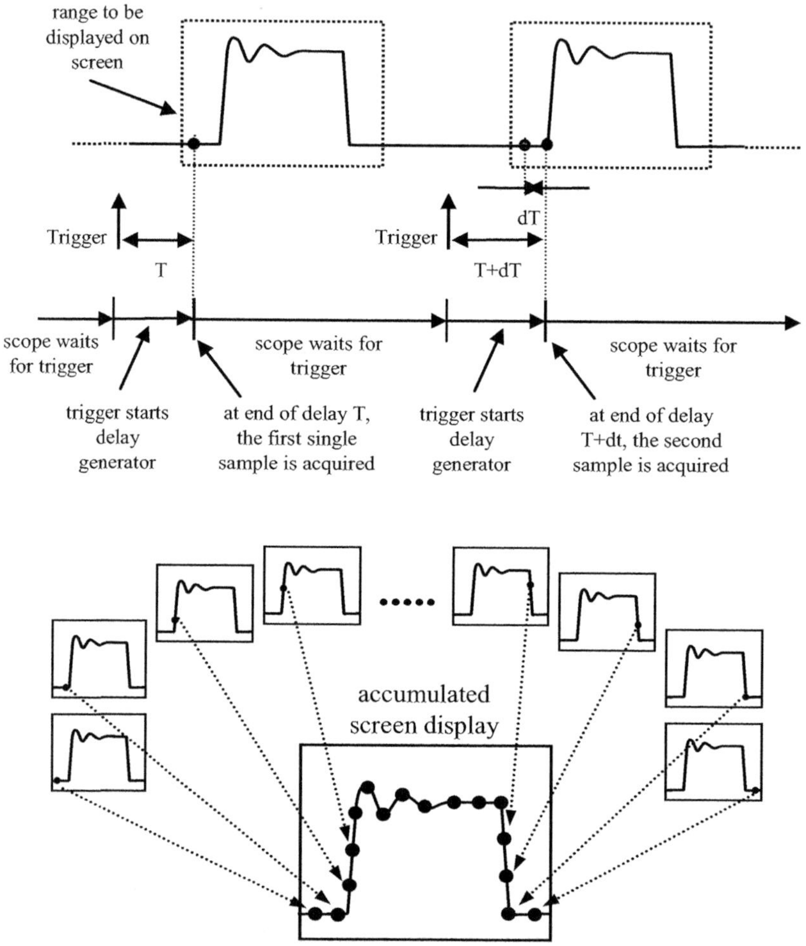

Figure 23: Principle of operation for equivalent-time sampling.

When the trigger arrives, the scope needs a short time (around 20 ns is a typical number) to set up the sampler which then acquires a single data

[30] Note that it is not absolutely necessary that the signal is *periodic*; it only has to repeat with a fixed relation between trigger and signal.

point. Then the scope waits until the next trigger arrives, the trigger is delayed a tiny bit longer and another sample is taken. Repeating this process many times yields a series of samples at increasing *equivalent times* along the waveform, which is then displayed on the screen.

Since only a single sample is taken at once, all effort can be put into this, and so the sampler can achieve much higher resolution (up to 14 bits today compared to 8 for real-time scopes), which compensates somewhat for the lack of an amplifier for small signals, and provides unsurpassed resolution (and low noise) for larger signals.[31] The absence of the input amplifier also means that zooming into the waveform (increasing the voltage resolution) merely changes the *displayed* voltage range, but not the *actual* resolution (that is fixed by the total dynamic range and the number of sampler bits). So – unlike on real-time scope – the fact that the waveform exceeds the displayed range does not necessarily mean we are overdriving and saturating the input.

The *equivalent* sampling rate – i.e. the difference in delay from the trigger from one sample to the next – is only limited by the accuracy with which the scope can generate this delay. Resolution well below 1 ps is standard on today's scope models (compared to at least 25 ps and more even on the fastest real-time sampling scopes). On the other hand the *true* sampling rate – the number of samples acquired per second – is rather low, even the fastest such scopes don't exceed a few MSamples/s (most are in the kSamples/sec range). This becomes very visible if many waveforms have to be acquired (e.g. for averaging) or if the trigger does not comes very often (e.g. a frame trigger in a very long data pattern) – the lower of the maximum sample rate and the trigger rate determines the number of samples taken per second! Out of that reason the record length on equivalent-time sampling scopes is usually limited to a few 1000 points at best.

Another detail is that the input impedance into an equivalent-time sampling scope (and thus the load put onto the circuit under test) is virtually always 50 Ohm. True, one could use an active scope probe in front, but that – the probe being a bandwidth-limited amplifier circuit – would destroy the major advantage of such a scope – large bandwidth. The only exception may be when one needs to probe a differential signal since not every such scope on the market offers true differential inputs.

There is really no good technical reason to use an equivalent-time sampling scope for applications where the timing resolution, bandwidth, and timing accuracy of a real-time scope are sufficient, but equivalent-time sampling scopes are among the very few instruments suitable for today's highest data rate signals (above maybe 6 Gb/s) and for maximum-accuracy

[31] We can easily reduce the amplitude of signals that are too large with a passive high-bandwidth attenuator.

measurements. Among the tradeoffs are acquisition speed and flexibility. The fact that it takes some time for the scope to strobe the sampler after the trigger has been received means that just like an analog scope (and unlike a digital real-time scope) an equivalent-time sampling scope cannot directly show what happens at or before the trigger instant, unless a delay line is used to delay the data signal by more than the minimum data delay – but such a delay line will degrade the path bandwidth and thus negate at least part of the reason for using an equivalent-time sampling scope in the first place.

Because the market for such ultra-high bandwidth and accuracy instruments is still rather limited, and because of all the difficulties of designing such an instrument that exceeds the performance of the best real-time scopes, there are only very few test equipment vendors active in this area.

One final reason for using an equivalent-time sampling scope is that they are relatively inexpensive – they tend to cost only half or less for the same (or higher) bandwidth compared to a real-time sampling scope or a BERT box.

2.4 Time Stampers

Time stampers are fundamentally different from oscilloscopes in one important respect: Oscilloscopes are basically fast voltmeters, i.e. they provide the instantaneous voltage at given instants in time, or in other words, they measure the voltage when the time reaches a certain value. On the other hand time stampers do exactly the opposite: They measure the timing whenever the input voltage crosses a certain user-adjustable threshold. While the oscilloscope stores a series of such voltage-versus-timing measurements (either directly on the screen for an analog scope, or in memory in the case of digital scopes), the time stamper stores the sequence of timing numbers in its memory for subsequent analysis.

Figure 24 depicts a typical block diagram of such a time stamper (as usual, the architecture of a specific implementation may differ, but the example is intended to show the relevant principles). It consists of two slope-sensitive comparators (so one can choose to look at rising or falling edges only, respectively), a fast timing system consisting of a highly stable master clock and an interpolator, and storage memory to keep the results. The master clock – running maybe at 10 or 100 MHz – provides a stable time base, while the interpolator enables fine timing resolution (down to ps).

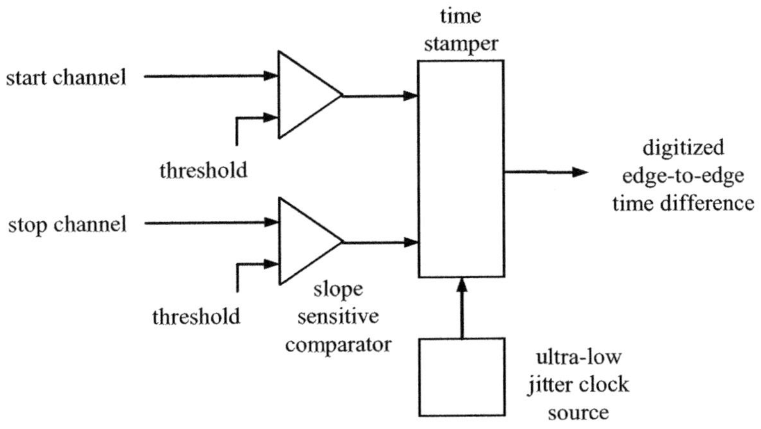

Figure 24: Block diagram for a typical time stamper.

An obvious question that arises is why *two* identical comparators are necessary: The comparators employed as well as the electronics to transfer the results into memory have only limited re-fire rates, i.e. after the comparator triggers it takes a while before it can trigger again. Those dead-times are often of the order of a few ns or more, much too long to be able to capture two subsequent edges of a fast data stream. Having two comparators allows having the second one waiting until the first has fired, so one can capture arbitrarily small delays between events.

There is a variety of timing numbers that we can get this way:

- Signal period: One comparator triggers on a certain edge polarity (rising or falling edge, respectively), the second triggers on the following edge of the same polarity.
- Pulse width: One comparator triggers on a certain edge polarity (e.g. rising edge), the second triggers on the following edge of the opposite polarity (falling in this case).
- Rise times: One comparator triggers on a certain edge polarity at the low threshold (e.g. rising edge, 10% of swing), the second triggers on the same edge but at a different threshold (e.g. 90%, so the timing difference gives the 10/90 rise time).
- Duty cycle: This is simply a combination of period and pulse width measurements.
- Phase noise (N-period jitter): Similar to signal period measurement, but now we measure the jitter between an edge and one several (or many) cycles later. A sweep of the number of cycles (e.g. ranging from one

cycle up to hundreds or even thousands), for each setting acquiring the N-period jitter statistics, gives the jitter trends versus the time delay or, in frequency domain, the phase jitter vs. frequency (inverse of time delay). Unlike an equivalent-time sampling scope a time stamper can handle jitter exceeding a bit period because it does not acquire after a specified time, but rather after a specified number of *edges* and so is immune to large *cumulative* edge movements.

Statistics over many subsequent such measurements can give values for timing jitter, as we will discuss later in this book.

As long as we are only concerned with *timing* measurements, time stampers have the advantage of acquiring exactly the data that we want and nothing more, while scopes always gather the full two-dimensional waveform but in principle we are only interested in a single point (or a few) on this curve, namely the threshold crossings. What's more, in order to achieve reasonable measurement resolution oscilloscopes must have very high sample rates, which complicates their design and increases the amount of (for pure timing measurements unnecessary) data acquired, slowing down data acquisition and processing. Of course scopes can do many things time stampers can't (e.g. signal noise, ringing, and overshoot measurements), because they provide more information (two-dimensional curves vs. one-dimensional timing values).

It seems thus fair to say that for troubleshooting and for comprehensive measurements scopes can't be beaten, but when it comes to speed of acquisition for pure timing measurements time stampers have a well-established niche. For that reason one finds them often as part of a large-scale digital production tester where speed of test and automation of the acquisition process takes priority. If they are standalone units, they are often called *time interval analyzers* (TIAs) but the functionality is largely the same. Leading-edge time stampers can acquire up to maybe 100000 timing values per second with minimal overhead, and they also offer true differential inputs, so measurements can be made on differential signals as well.

Apart from the fact that they do not measure voltage but timing, the design of time stampers puts an additional limitation on the capture of subsequent edges in a fast data stream: Due to the relatively slow refire rate a time stamper with two comparators can only acquire two edge timings, then there is a rather long break (at least a few ns, if not µs) before more edges can be captured. This runs the risk of missing any medium and long range effects. Some recent models of time stampers seek to improve the situation by using up to 10 comparators, but even this limits us to 10 subsequent edges. In that respect they are well inferior to real-time sampling scopes (but not equivalent-time sampling scopes), which in a single shot can continuously capture waveforms containing thousands if not millions of edges. As a consequence time stampers have to make certain assumptions and some modeling if they want to determine jitter numbers decompose jitter into its components – in the next chapter, about jitter and jitter measurements, we will go into some more details about that.

A final limitation is that even the most recent time stamper models max out at an analog bandwidth of around 3 GHz. Remembering that to do meaningful measurements our bandwidth must be *at least* three time higher than the highest frequency of interest, this makes them usable for data rates of at best 2 Gb/s (for double-data-rate signaling) or clocks running at 1 GHz – and today's fast serial busses already exceed this range. Time stamper cards integrated into production testers normally don't even reach those bandwidths; they rarely even attain 1 GHz bandwidth.

2.5 Bit Error Rate Testers

In principle, when transmitting *digital* signals, all we *really* care about is if the digital information (the ones and zeros) makes it from the sender to the receiver without error, where an error would be a one received as a zero, or a zero received as a one. Maintaining signal integrity and clean waveforms is just a means to an end – the receiver shall be able to distinguish zeros from ones. No transmission is absolutely perfect and error free (see the discussion about random jitter later in this book), so usually all we can do is guarantee a certain maximum *bit error rate (BER)*. This BER is the number of "broken" bits (bits that get received incorrectly) to the total number of bits transmitted. This is where bit error rate testers (BERTs) come in.

BERTs basically consist of a fast data source as well as a fast receiver (level comparator, often with programmable threshold). The principle BERT setup is shown in Figure 25. Since BERTs originate from serial data transmission schemes, most of them have only serial drivers and receivers, but in principle nothing prevents us from building a BERT box with a parallel data source. What a BERT does is create a data stream, send it to the system under test, receive the transmitted data stream, compare it to the original data and count the number of failing bits – which gives the BER. Very often the data stream is some sort of pseudo-random bit stream (PRBS, more details to that in the next chapter) that assures all different possible bit sequences up to a certain maximum length are present in the data stream, but depending on the application the stream may also be any type of user-defined bit sequence, for which purpose most BERTs have a large built-in linear pattern memory.

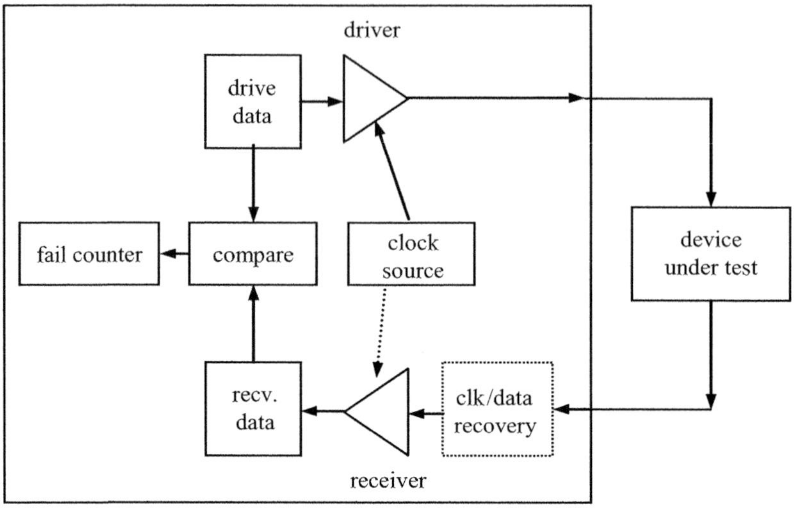

Figure 25: Block diagram of a bit error rate tester (BERT).

In the final application the receiver's strobe is likely to be placed at the center of the bit period (also called the *unit interval, UI*), but a BERT box allows us to move the strobe around. Without going into too much detail (we will dig into it more in a later when we look at jitter) it is clear that if we move the strobe closer and closer to the ideal position of the transitions, due to timing jitter the signal will more and more likely transition at the wrong side of the strobe, so the BER will increase. In that way a plot of the measured BER versus the respective strobe positions yields statistical information about the timing jitter of the signal averaged over many transitions, and thus – indirectly – edge timing information.

One big advantage of a BERT is that by design it captures pass-fail information from *every* bit even of a long data stream, so unlike e.g. time stampers or equivalent-time sampling scopes it has no dead times where it may miss failures, and it does so very fast (with the transmission data rate). So even if some effect only appears very rarely, e.g. for a very specific combination of bits, it will capture it as long as the bit combination is contained in the pattern.

On the other hand, a BERT only gives statistical jitter information, but no direct timing data. E.g. if a certain bit failed, we don't know how the failure looked like, not even if it just barely missed the right timing or if it was completely off. Also, while a single measurement (running the pattern once for a specific strobe position) may be relatively fast, the acquisition time for a full scan of the strobe timing over a bit period soon becomes excessively long if we demand good timing resolution (i.e. small increments of the strobe timing position). In addition, the receiver being a simple single-threshold comparator, a BERT cannot give us much information in the voltage domain.

2.6 Digital Testers

Digital testers, be it for large-scale production test or for bench-top characterization work, are geared towards the *functional* test of devices with a large number of digital data pins. The basic emphasis in those machines is thus overall on sending and receiving digital bits rather than measuring analog waveforms and timings. Since for digital tests it is sufficient to determine if a signal is higher or lower than some threshold, but the absolute value is of no further importance at least as far as functional test is concerned, they employ so-called *comparators* to measure the incoming signals – identical to a BERT box with the difference that the latter rarely has more than one or maybe a few channels.

From a user's point of view one can see those comparators as simple switches that will yield "high" whenever the incoming signal exceeds the threshold and low whenever the signal is below the threshold (in many testers the comparator – from the user's point of view – seems to have *two* thresholds, where "low" means "lower than threshold 1" and high means "higher than threshold 2", but internally this is realized as just two single-threshold comparators in parallel). The threshold itself is user selectable (programmable), although very often only when no pattern is running. Also programmable are the timing during a test pattern run when the comparator shall *strobe* (i.e. transfer the comparison result to the digital capture memory). This is usually possible once within a device cycle period, e.g. every 500 ps when the test data rate is 2 Gb/s, and the results of those comparisons are captured in real time.

Such an input circuit is simple compared to the sophisticated sampling hardware that a scope requires, so it is possible (financially as well as complexity-wise) to have one comparator on each of the many channels. The comparators can be single-ended or differential depending on the design of the tester's pin electronics.

If we want to relate comparators to scopes (since they, too, measure voltages at specific instances in time), we can regard them as 1-bit A/D converters with adjustable threshold. A single bit is enough to distinguish between "higher than threshold" and "lower than threshold", but not more. The real-time strobe rate is sufficient to capture every single bit in a data stream, but much too slow to accurately resolve transition timings or detailed waveforms (at least in real time), last but not least because digital testers normally lack the sophisticated interpolation algorithms of oscilloscopes. But in the last chapter of this book we will see how one can nevertheless accurately trace analog waveforms (i.e. emulate the functionality of an oscilloscope) and measure edge timing and jitter.

Since the input side of a comparator is just an analog electronic circuit (which does not care if its output gets sampled with 1 bit resolution or with 16 bit), it is of course subject to the same considerations regarding analog bandwidth and rise time as any oscilloscope. Because the main emphasis in the comparator design is usually on small, simple, inexpensive circuitry (after all, the tester designer needs to integrate hundreds if not thousands of those into his tester), not optimum waveform fidelity, the achievable accuracy is limited.

2.7 Spectrum Analyzers

A spectrum analyzers is quite a departure from all other instruments that we have looked ad, insofar as it does not measure the signal in the time domain (i.e. signal amplitude vs. time) but rather in the frequency domain (i.e. signal amplitude vs. frequency). Another way to see this is that it does a decomposition of the signal into a Fourier spectrum of sine waves of different frequencies and phases, and shows the amplitude (power, to be exact) of each of those sine components. Of course this spectrum is related to the signal in time through Fourier transformation, so in principle a spectrum analyzer can yield the same fundamental information as an oscilloscope.

Traditional spectrum analyzers are built as shown in Figure 26: The signal passes a very narrow-band bandpass filter[32] that filters out everything except a small band around its center frequency. Those filters are multi-pole filters (4 poles is common) and thus have a much steeper roll-off away from their center than a simple Gaussian filter. The pass-band can be as narrow as just a few kHz or less. The center frequency can be swept over a certain range with high resolution, and the spectrum analyzer shows the amount of signal that goes through for each frequency – resulting in the frequency (Fourier) spectrum of the signal.

[32] As shown in Figure 26, the actual design employs a mixer/downconverter that mixes the incoming signal with a locally generated swept-frequency signal. The mixing product contains both sum and difference components of difference and local frequency. A constant-frequency bandpass filter extracts the difference frequency. While technically very different, this whole contraption is functionally equivalent to a swept-frequency bandpass filter, but makes it easier to achieve the required filter characteristics because all subsequent stages only have to deal with constant-frequency signals.

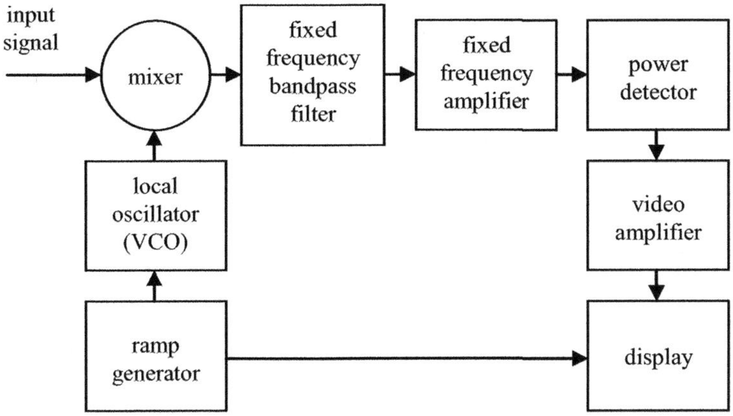

Figure 26: Simplified block diagram of a spectrum analyzer.

This architecture allows us to achieve huge frequency ranges (good high-end spectrum analyzers can cover a range from just a few Hz all way up to over 100 GHz) with amazing selectivity (filter bandwidth). This is difficult to match with an oscilloscope: On the low end a frequency of a few Hz corresponds to acquisition sets that must cover close to a second in time. High maximum frequency means the scope would have to do this with a very high sample rate. Both together result in unrealistically large data sets – no currently available high-end real-time scope has a capture memory of more than a few 10 MSamples, which limits the time span at maximum sampling rate (about 40 GSamples/sec) to less than a millisecond. At the same time the fact that a spectrum analyzer basically does a steady-state (DC) measurement of an amplitude enables unmatched signal-to-noise ratios (well over 100 dB is common) and thus measurement accuracy.

Unfortunately the loss of phase information (as only the amplitude is retained when passing through the filter chain) prevents us from fully reconstructing the original signal in the time domain, as already illustrated in Figure 1 (section 1.1). There is one class of spectrum analyzers that acquires the signal like a real-time sampling scope and then applies real-time FFT (Fast Fourier Transformation) to it. This preserves the phase, but on the other hand the acquisition architecture is the same as for an oscilloscope and thus holds no advantage over them regarding maximum or minimum frequency. Also their signal-to-noise ratio (SNR) is limited to maybe 40 – 60 dB, compared to 100 to 130 dB for "real" spectrum analyzers. Actually many of the higher-end real-time sampling oscilloscopes offer real-time FFT as a possible display mode.

Spectrum analyzers are great tools to track down and characterize periodic features of the signal (e.g. periodic jitter), and can also be useful and very accurate (because of their superior signal-to-noise ratio) in cases where the phase information is not absolutely needed – one case being random jitter measurements. We will see more about that later in this book.

3. KEY INSTRUMENT PARAMETERS

3.1 Analog Bandwidth

When looking at a scope (or similar measurement device), we see that the first step in acquiring the signal is to deliver it to the digitizer (analog-to-digital converter, ADC). The part before the ADC is considered the *analog* portion of the signal path, while everything in the signal chain behind the ADC is considered the *digital* portion.

Being purely analog, any distortion on the signal will enter into the displayed result; as usual, we are mostly interested into the low-pass filter behavior that will limit the maximum frequency (minimum rise time) that can be delivered to the sampler – if the signal (or better: its high-frequency components) does not reach the sampler, then the best ADC is of no help – the signal is irreversibly lost.[33]

In the previous chapter we have learned that rise time and bandwidth are connected by the simple formula

$$T_r = \frac{k}{BW_{-3dB}}, \quad k \approx 0.33 \ldots 0.6 \qquad (24)$$

Small k's indicate a rather smooth drop-off of the response with frequency, while higher numbers correspond to a faster, "brick-wall"-like drop-off. Examples for the former are Gaussian and simple exponential (single-pole) filters, while filter types like Chebyshev and Butterworth are representative of the latter category. Figure 27 shows some typical filter response curves.

Many commonly used formulas and rules of thumb assume Gaussian filters – one example being the addition of rise times (or inversely the calculation of the true input rise time when the system rise time is known). Unfortunately, modern digital oscilloscopes rarely behave Gaussian, but instead have a steeper drop-off beyond the 3 dB point. One reason is that in

[33] As long as at least some portion of these components is preserved one can theoretically re-create the full signal numerically through digital signal processing (see the last chapter in this book), but even this method reaches its limits rather soon with increasing attenuation.

order to avoid aliasing[34] the sampler must not get any frequency components that exceed half the sample rate, so the analog front-end strives to filter those components out. The second reason is that a Gaussian response – with its gradual drop-off not only beyond but also before the 3 dB point – would mean that we have considerable attenuation even well below the 3 dB bandwidth limit, which is definitely not what we want because it introduces large amplitude errors even for moderate rise times.

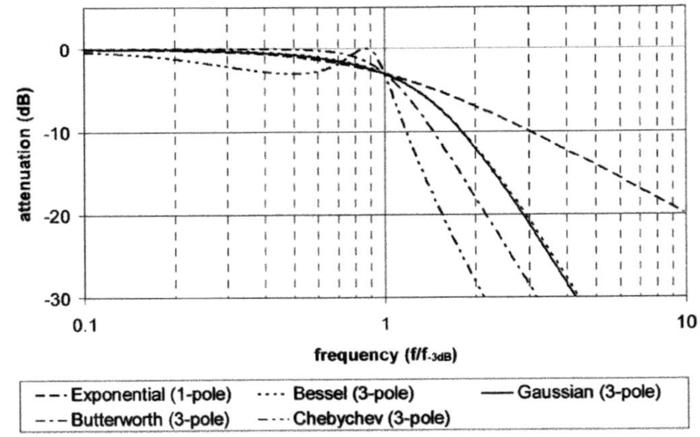

Figure 27: Filter responses (transmission loss vs. frequency) for several different commonly used filter types.

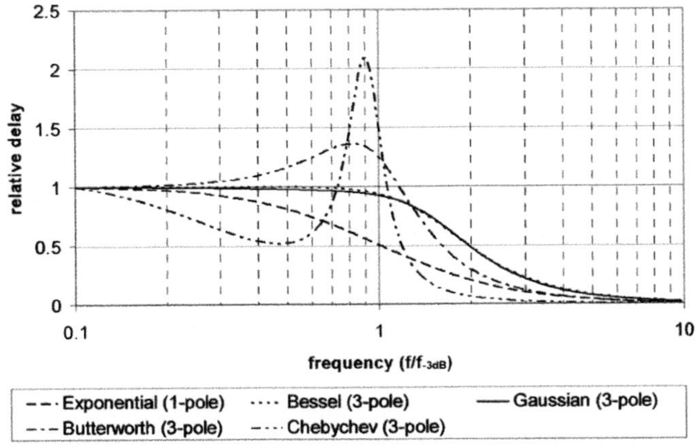

Figure 28: Filter delay vs. frequency (i.e. dispersion) for several different commonly used filter types.

[34] Aliasing means high frequency components of the signal get "folded back" into the range below the sampling bandwidth, causing distortion and artifacts in the sampled waveform.

At first glance the solution to both problems seems to be to implement one of the "brick-wall" filter types. Unfortunately, while they look tempting in frequency domain (where they have indeed wide application), such filters don't behave nicely in time domain: they exhibit large ringing that takes a long time to settle, as shown in Figure 29. They also have a fair amount of dispersion, meaning signals of different frequencies experience different delays through the filter (see Figure 28). An edge (consisting of a broad spectrum of frequency components) will get "washed out" because some components take longer than others to traverse the filter; this will also cause edges with different rise times to experience different delays, causing timing errors. Thus the scope designers have to make some tradeoffs between high-frequency filtering, flat attenuation curve below the 3 dB bandwidth, minimum dispersion (timing delay change with frequency), and clean time-domain response. The end result is a filter with k of around 0.4 – a bit closer to a Gaussian filter than to Chebyshev of Butterworth.

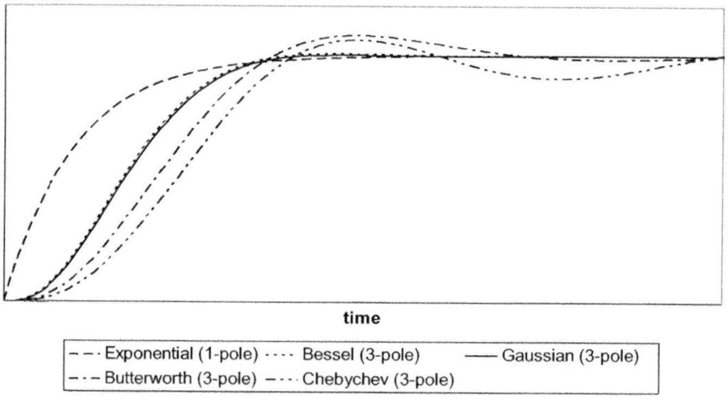

Figure 29: Filter responses in the time domain for several different commonly used filter types.

This deviation from Gaussian behavior means we have to be somewhat careful when applying our rules of thumb; they are *approximately* valid and work as long as the corrections are not too large. E.g. let's consider a scope with 1 GHz analog bandwidth, which measures a signal edge; assume that the 10/90 rise time we see on the screen is 1.1 ns. What is the true rise time of the signal going into the scope?

We know that rise times add up as squares (exactly valid only as long as all edges and filters are Gaussian), thus

$$T_{r,true} \approx \sqrt{T^2_{r,meas} - T^2_{r,scope}} \approx \sqrt{T^2_{r,meas} - \left(\frac{0.4}{BW_{r,scope}}\right)^2} \qquad (25)$$

Putting in the numbers we get

$$T_{r,true} \approx \sqrt{(1.1\,\text{ns})^2 - \left(\frac{0.4}{1\,\text{GHz}}\right)^2} \approx 1.025\,\text{ns}\,.$$

In other words the calculated correction is of the order of 7%, which qualifies as "small", and we can assume that even with the non-Gaussian behavior of the scope the result is reasonably accurate. But don't try this in cases where the scope rise time is of the order of (or larger than) the signal rise time!

3.2 Digital Bandwidth; Nyquist Theorem

A second limiting factor in the acquisition process is the sample rate. Keep in mind that instrument "bandwidth" (or "rise time") is simply a measure of how fast the instrument's acquisition system can react to a signal level that changes in time. It does not make any restrictions as to *what* exactly causes this limitation in reaction time.

If a signal makes a sudden jump between one sample and the next (e.g. at one sampling point it is still low, and at the next sampling instant it is already high), then the scope has no way of knowing what exactly happened between the two sampling instants – it only knows the original and the final state. Any signal that rises faster that the sample interval will result in the same sampling result, no matter if it rises within $1/10^{th}$ of the interval or if it takes the full interval to rise. So the scope has to make some assumption about the interval between them, and the simplest approach is to assume that the signal just rose linearly between those two points. This is illustrated in Figure 30(a). There we can also see that there is an even worse case where it looks to the scope as if the signal actually needed *two* intervals to rise. So in other words, from Figure 30(b) we can deduce that without any further data processing (interpolation) the limited sample rate is equivalent to an effective rise time (or bandwidth) limitation somewhere between 0.8 and 1.6 times the sample interval:

$$T_{r,dig} = (0.8\ldots1.6)\times\Delta T_{sample}\,, \tag{26}$$

or to a so-called *digital bandwidth* of approximately

$$BW_{dig} \approx \frac{0.4}{T_{r,dig}}\,. \tag{27}$$

This is a serious limitation for any real-time sampling scope where we cannot increase the sample rate above a certain limit (40 GSamples/sec for the best scopes available today). Another conclusion from above formula is that our sample rate must be at least twice as high as the highest frequency component of interest. All this is of much less concern on equivalent-time sampling scopes because there the (equivalent) sample rate can be made almost arbitrarily small (well below 1 ps on the best available scopes, corresponding to a *digital* bandwidth of over 400 GHz, which is one of the main reasons to use this type of instrument).

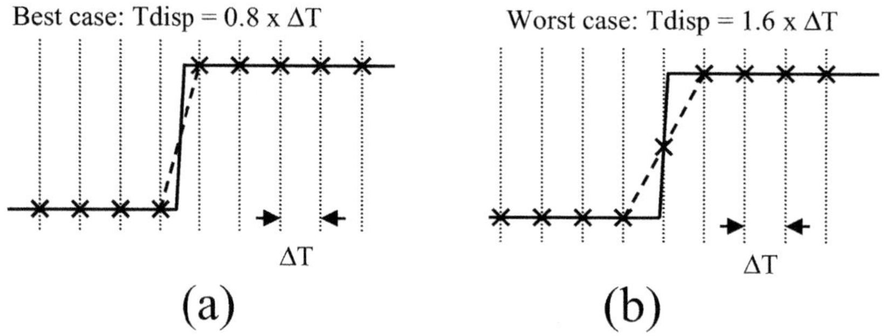

Figure 30: Rise time limitations caused by limitations in the sample rate (digital bandwidth): (a) best case, (b) worst case.

As a matter of fact above considerations were a bit "cavalier" on the impact of frequency components in the order of (or higher) than the sample rate. Mathematical theory of sampling – the well-known Nyquist theorem – gives a bit more stringent conditions: According to this theorem the sample rate must be *more than twice* the highest frequency component in the signal (the *Nyquist* frequency).

The theorem stems from an effect that is called *aliasing*. This effect is easily illustrated in Figure 31: It shows a sine wave of 1 GHz, sampled at a rate of 1 GSamples/sec and 1.5 GSamples/sec, respectively. Evidently the sample points at 1 GSamples/sec always fall on the same spot on the sine wave, and the reconstructed signal has constant level. The slightly faster sample rate (1.5 GSamples/sec) falls on different places on the curve, so the reconstructed signal is not a steady DC signal but varies in time, but with an apparent frequency of only 0.5 GHz. Those two cases illustrate a general principle that is dealt with in detail in the mathematical theory of Fourier transformation: Let's assume a signal with some frequency spectrum (harmonics) that is sampled at some sample rate. If we now reconstruct the signal from the sampled points (as done in the upper half of Figure 31), we will find that the signal has been distorted (through the low-pass filter effect). But what's even worse, the frequency components exceeding half the sample rate are not simply gone, they get "folded back" (mirrored) into the frequency spectrum, the mirror being the Nyquist frequency (equal to half the sample rate): In our case the Nyquist frequency for 1 GSamples/sec is 0.5 GHz, so the 1 GHz signal mirrored into a 0 GHz signal, i.e. a DC level, just as we see in Figure 31. For the case of the 1.5 GSamples (Nyquist frequency 0.75 GHz), the 1 GHz signal is mirrored back into a 0.5 GHz signal. This is also indicated in Figure 31. What makes this effect so devilish is that after sampling, if we take the 1.5 GSample case, a mirrored (aliased) 1 GHz signal becomes absolutely and perfectly indistinguishable from a "true" 0.5 GHz signal – they both look like 0.5 GHz. No amount of digital signal processing can reverse this!

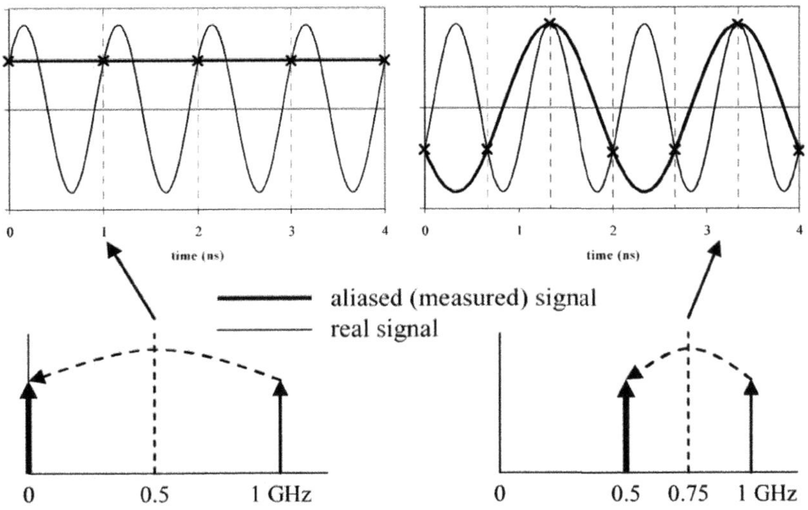

Figure 31: Aliasing caused by undersampling (insufficient sampling rate): The signal frequency component gets mirrored at the Nyquist frequency (identical to half the sampling rate). Upper half: time domain representation, lower half: frequency domain representation.

As a consequence, in order to avoid these effects one must thus make sure that no frequency component higher that half the sample rate gets to the sampler in the first place. In other words, some *analog* signal conditioning is indispensable before the conversion of the signal into digital numbers. Oscilloscope designs do that by matching the analog bandwidth to the available digital bandwidth, so the analog portion provides the necessary filtering. It should also be noted that the limit of 2 (sample rate equal to twice the Nyquist frequency) is not attainable in practice. Realistic numbers for good oscilloscopes are in the range of 2.5 to 3. And indeed, if we take a look at different oscilloscopes, we will usually find out that the sample rate is approximately three times the analog bandwidth or more – and now we know why![35]

3.3 Time Interval Errors, Time Base Stability

Whenever we talk about timing, we mean the position in time *relative to some reference*. There just isn't anything like an absolute time (like "noon" or 3:25pm) – our references tend to be transitions on some trigger or clock channel, or maybe some specific edge in our data stream. As a result, what we really measure and display are really time *intervals* between two events (reference event and event of interest). Thus the important figure of merit for any instrument is how accurate it can measure those intervals.

Every oscilloscope, BERT box, time interval analyzer etc. acquires the signal based on some internal timing reference (often called the *time base*). Any inaccuracy in this time base will show up directly as an inaccuracy in the acquired timing information. From a high-level point of view we can separate those inaccuracies into two major types of errors: random and deterministic. Random errors will be different every time we repeat the measurement. Deterministic errors will always be the same as long as the test conditions did not change, and in the present context they are commonly referred to as *time base nonlinearity*[36]. We can compare this to measuring distances with a ruler that has slightly inaccurate markings. Those inaccuracies will show up as deterministic measurement errors, e.g. when we measure the length of an object that in reality is 1 m long, we may always get 1.01 because the 1 m marker is off. But at the same time our limitations in being able to see the exact marking may result in readings of 1.011, 1.008, 1.013 etc. if we repeat the measurement several times – these are random errors.

Achieving maximum time base accuracy is of course of utmost important to us, but it virtually always become more difficult the longer the intervals become. This is why most manufacturers specify the time base of their instruments as something like time base error = A + B × interval. Apart from choosing an instrument with good specifications for time base accuracy;

[35] One exception are lower-end oscilloscopes that sometimes offer excessive sampling rate compared to their analog bandwidth. This is because in the range they are working digital performance is cheap and oversampling eases the requirements on signal processing (interpolation), making for a cheaper overall design.

[36] We should note that it can depend on the specific measurement setup and the specific conditions if a certain deterministic *instrument* error shows up as (pseudo-)random *measurement* error or as a deterministic (constant for constant conditions) error. Also, in jitter analysis "deterministic" is usually assumed to stand for "non-Gaussian statistical distribution", while here we use it to denote the fact that the error will not change if we repeat the measurement.

special measurement methods can reduce the time base error even further. In the last chapter of this book we will look into this more closely (for the case of equivalent-time sampling scopes).

4. PROBES

4.1 The Ideal Voltage Probe

When dealing with high-speed digital signals, virtually the only probe types in use are voltage probes[37]. The probe enables us to connect our measurement instrument to the system under test. In this function we would of course like it to be as close to invisible as possible to our signal, while enabling reliable access to the probe points in our system. In other words, an ideal probe should have the following characteristics:

- It has zero rise time (i.e. infinite bandwidth).
- It does not distort the measured signal, i.e. it has zero (or at least constant) losses over the whole frequency range, and its time delay (group delay) is constant over the whole frequency range.
- It has infinite amplitude (voltage) range.
- It does not add noise to the signal.
- It puts zero load on the system under test, i.e. we won't influence the system behavior when probing it.
- Its size matches the available probing possibility (which for high-speed applications and modern fast devices usually means it has to be very small).
- It is mechanically rugged and reliable.
- Last but not least, it should cost as little as possible.

[37] We deliberately limit ourselves here to electrical signals. It is true that some of the highest-speed transmission systems use optical (light) signals, but no oscilloscope or tester can deal with such optical signals anyway. They all have to get translated into an analog voltage signal by the input stage first (in principle just a fast photo diode or photo transistor). Actually there are a few higher-speed current probes available, but the achievable bandwidth is much lower than for voltage probes – a few 100 MHz at best for the former compared with over 10 GHz for the latter. What's more, on a controlled-impedance transmission line (e.g. 50 Ω) in absence of reflections (i.e. matched termination at the end) there is always a direct correspondence between voltage and current, so measuring the voltage automatically provides the current profile as well.

Unfortunately, no single real-world probe can fulfill all those requirements perfectly at the same time, but it is often possible to get very close to ideal for a few of them. Every probe type we are going to look at in the following sections represents a compromise between all those competing goals.

The first major choice that we face is between active and passive probes.

4.2 Passive Probes

Like the name implies, a passive probe consists entirely of passive elements (resistors, capacitors, inductors, cables, etc.). As such, lacking any means of amplification, it can never provide more than the original signal amplitude to the measurement instrument. Since all the energy has to come from the original signal, the main tradeoff here is between measurement signal size and the load on the system under test[38]. On the upside, there is no hard limit for the maximum voltage such a probe can handle, the circuitry is simple, and it does not need any power supply, all this making it a very inexpensive solution. Being purely passive, it also will not add any random jitter or noise to the signal (at least as long as it does not pick up external fields, but we will discuss that later).

A passive probe can be as simple as a piece of cable. The effective load on the system depends strongly on how the path is terminated in the oscilloscope[39]: Until several years ago, when the maximum frequencies in digital systems were low (a few MHz at best) and many device drivers were not designed to drive a 50 Ω load, the scope input was high impedance, and everybody assumed the load to be the purely capacitive (using the total capacitance of the cable). But as we already know this completely neglects the distributed characteristics of this cable capacitance as well as its inductance, and as a consequence delay, reflections and ringing. Once the signal bandwidth becomes high enough, transient effects (reflections at the unterminated end) would completely dominate the picture. Overall, at today's data rates the scope always needs to terminate the line to 50 Ω to avoid reflections, and the load then becomes a constant 50 Ω (and *not* capacitive, except for small parasitics that we will talk about later). This may be fine in a device characterization setting where all the driver needs to drive is the scope, but it will wreak havoc if we attach such a "probe" to the

[38] The larger the load, the more the system behavior will be influenced by the presence of the probe, which is normally not a desired property.

[39] For simplicity we will name only oscilloscopes, but all that is said implicitly applies to any other instrument (BERT, spectrum analyzer, production tester) as well.

middle of a data bus –chances are the additional load (and the impedance mismatch created by it) will make the data transmission fail immediately, not even considering the fact that this will not show us what the signal really looks like without the probe attached.

One solution is to increase the probe impedance by adding a resistor at the probe tip so it acts as a voltage divider (made up by the resistor and by the characteristic line impedance). Ratios of 1:10 or 1:20 are typical (i.e. 500 Ω or 1 kΩ load impedance, respectively[40]). In the simplest incarnation the divider is a single higher-impedance resistor in combination with the 50 Ω cable/scope impedance. While this does reduce the load on the system under test, it also greatly reduces the signal amplitude transmitted to the oscilloscope, so the scope's noise has a larger impact (reduced signal-to-noise ratio), which will make it difficult to observe and measure very small signals. What's more, large impedances make the probe extremely sensitive to any parasitic capacitance (remember that the time constant is $R \times C$ and the bandwidth is inversely proportional to that). The resistor needs to be placed right into the probe tip; otherwise the transmission line stub from the probe tip to the first resistor will cause reflections and ringing. The residual stub plus some unavoidable fringe fields at the tip inevitably add some parasitic capacitance to the load presented by the probe – the smaller this capacitance is, the better (more to that later). This capacitance is usually given in the manufacturer's specifications for the probe.

It is easy to calculate that the impedance mismatch created by an additional 500 Ω load placed on a 50 Ω transmission line will cause approximately 5% reflection and will reduce the signal amplitude transmitted to the receiver by almost 5%. It depends on the particular case if such an impact is acceptable or not.

That said, passive voltage-divider probes can perform well up to several GHz, and their simplicity makes them a valuable tool for low- and midrange applications. If the bandwidth requirement is not very high, it is easy to build such a probe out of a small surface mount resistor soldered onto the tip of an SMA connector.

4.3 Active Probes

Putting an amplifier – i.e. an active element – in the probe head is clearly a way to overcome the loading-vs.-signal-amplitude tradeoff. Well designed

[40] For 500 Ω probe impedance, the necessary series resistor is 450 Ω. In series with the characteristic line impedance of 50 Ω this adds up to the desired value, and makes for a 1:10 division ratio. Note that for a 50 Ω signal source this reduces the swing visible at the scope by only a factor close to 5, not 10 (because the load is now negligible, while a 50 Ω load reduces the driver swing by half).

integrated MOSFET amplifiers exhibit input impedances up to several hundred kΩ or more, virtually eliminating any static load on the measured system. Leading-edge probes achieve bandwidths exceeding 10 GHz. The main detractor is the parasitic capacitance (again the amplifier has to be placed right at the probe tip, but some minimal stub is unavoidable), which we will look at more closely shortly, but a good high-performance probe will have only a fraction of a pF.

Not surprisingly this improvement in performance comes at a price. Fast transistor circuits are very sensitive to overvoltage or electrostatic discharges – simply touching them with our finger when we are not well grounded can destroy them, and a good active probe is expensive (much more than a good passive probe). They need an external power supply (but this is often provided built into the oscilloscope). Being active circuits they will always add some noise and some random jitter to the signal, but this is usually more than compensated by the increased signal amplitude delivered to the oscilloscope. And finally, there is a clear tradeoff between electrical and mechanical performance – to minimize parasitics and to allow to probe today's narrow-pitch device packages and connectors, the tips of high-bandwidth probes (and the ground leads as well) must be small and correspondingly fragile.

4.4 Probe Effects on the Signal

As we have already mentioned in the beginning of this section about probes, some influence of the probe on the system under test (and consequently on the signal to be measured) is unavoidable. In this section we will investigate this in a bit more detail, with the goal of understanding typical problems and looking for possible remedies.

4.4.1 Basic Probe Model

Figure 32 displays a simple, very generic model of a probe. While it does not claim to be a very realistic or detailed model of any real probe, it does include all the necessary components needed to get a good general understanding of how a typical probe behaves in the system. We will have a closer look on each of those probe components in the following sections.

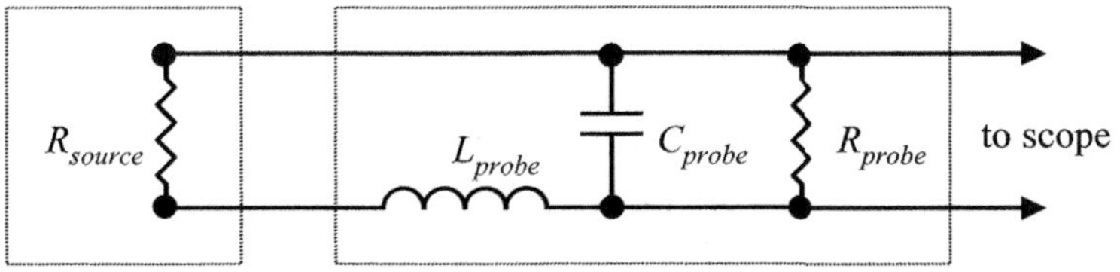

Figure 32: Basic, simplified model of an oscilloscope probe.

4.4.2 Probe Resistance

First, every probe has some static input impedance, represented by the ohmic resistance R_{probe}[41]. We have already seen that this resistance is rather low (usually less than a kΩ) for passive probes, and much higher (at least a few ten kΩ) for active probes. This load will reduce the signal amplitude in the system under test and may even potentially affect the circuit's operation. It will also create additional reflections when the probe point is in the middle of a signal line. Fortunately, at least for active probes, this load is too small to be of much importance in real life, unless the source impedance is very high (which is more of an issue with on-chip probing than with line drivers, since the latter have to be strong – i.e. low impedance – enough to drive a 50 Ω transmission line).

4.4.3 Parasitic Probe Capacitance

Of much more concern is the parasitic probe capacitance C_{probe}. It is the sum of the probe tip's fringe capacitance, the stub leading to the resistor or amplifier, and – for active probes – the input gate capacitance of the amplifier. As we all know the impedance of a capacitance C_{probe} is inversely proportional to the frequency *f*, namely

$$Z_C = \frac{1}{2 \times \pi \times f \times C_{probe}}. \tag{28}$$

For frequencies above a certain limit this shunt capacitance will have much lower impedance than the ohmic probe resistance, thus completely dominating the probe's total impedance. The effect is illustrated graphically in Figure 33: From that we see that the probe capacitance is a more

[41] The scope's input resistance – or more precisely, the characteristic impedance of the cable between the probe head and the scope – is of course in series with the impedance of a passive probe, but it is small compared to the probe resistance.

important factor for high-speed probe performance than its DC resistance: The smaller the capacitance, the higher the frequency until which the probe impedance stays sufficiently high (i.e. much higher than the signal source impedance) to allow meaningful measurements. Another – maybe surprising – result is that for frequencies in the higher GHz range (corresponding to rise times below a few hundred ps) a straight cable connection is actually superior to all "real" probes – the great-looking 1 MΩ DC resistance specification of one's expensive active probe is of no real use here!

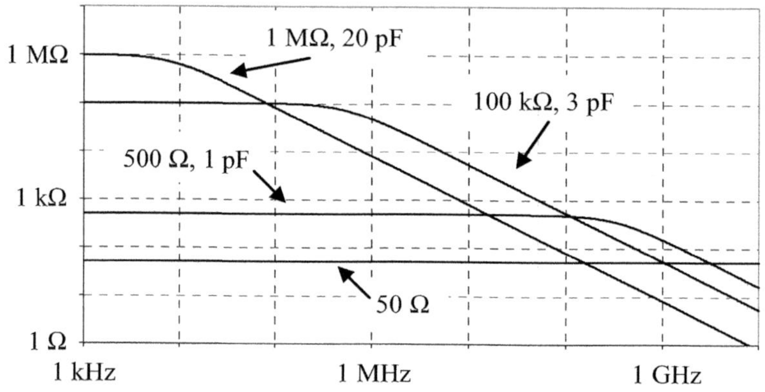

Figure 33: Effective input impedance vs. frequency for different probe types.

What's more, when probing in the middle of a signal line, the probe capacitance will cause a reflected spike going back to the driver, and it will degrade the rise time of the transmitted signal. As we know from our discussion about transmission lines and parasitics the rise time and bandwidth of such a low-pass R-C filter is $T_{10/90} \approx 2.2 \times C \times Z_0/2$ and $BW_{-3dB} \approx 0.35/T_{10/90}$, respectively, so apart from reflections and circuit loading the parasitic capacitance also limits the achievable probe bandwidth[42].

As an example, let's assume we have an active probe with a DC resistance of 100 kΩ and a parasitic capacitance of 1 pF, and we want to probe a signal with 100 ps rise time (the edge looks Gaussian). What can we expect to see? First, the rise time corresponds to a signal bandwidth of $0.33/0.1 = 3.3$ GHz. At this frequency the impedance of the capacitance is around 48 Ω, much lower than the DC resistance and closing in onto the

[42] Note that the factor ½ in the rise time formula is only valid if the probe point is either in the middle of the transmission line, or at the end when the line it is terminated with matched termination (in either case the two sides act together to form a 25 Ω Thevenin equivalent source). If probing at the – unterminated! – end of a line, the source impedance is 50 Ω which doubles the effective filter rise time and halves the bandwidth.

25 Ω effective source impedance, so we should expect to see some impact on the signal. The filter rise time is $2.2 \times 1\,\text{pF} \times 50\,\Omega/2 = 55\,\text{ps}$. The rise time seen by the probe amplifier (not counting the amplifier's own bandwidth limitations) will then be $\sqrt{100^2 + 55^2} = 114\,\text{ps}$. In addition attaching the probe to the transmission line will degrade the transmitted signal rise time to the same value, and also cause a reflected spike of approximately 20% of the amplitude, which is far from negligible[43]!

4.4.4 Parasitic Probe Inductance

The only component we haven't talked about yet is the probe inductance. It is caused by the current loop consisting of the signal path, the probe ground return path, and the return path through the system under test, as illustrated in Figure 34. As a general rule, the larger the enclosed area, the larger is the total inductance. This is one more reason why high-speed probes have to be so tiny.

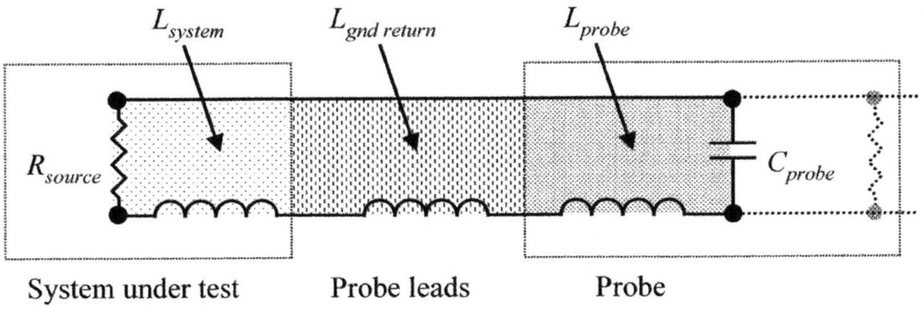

Figure 34: Current loop of system, probe, and ground return path, causing parasitic inductance.

In combination with the probe capacitance and the ohmic source resistance[44] this inductance forms a damped, resonant LCR circuit. Its resonance frequency – assuming small damping – is

$$f_{res} \approx \frac{1}{2 \times \pi \times \sqrt{L_{probe} \times C_{probe}}}. \tag{29}$$

[43] For the calculation remember that in the reflection formula the rise time is the 0%/100% rise time, which we can approximate as $T_{0/100} \approx 1.25 \times T_{10/90}$.

[44] In most cases the effective source resistance is much lower than the probe's DC resistance, so the former completely dominates. This is what we assume here.

How strongly damped (or how well oscillating) this circuit is depends on the so-called Q-factor

$$Q = \frac{\sqrt{\frac{L_{probe}}{C_{probe}}}}{R_{system}}. \tag{30}$$

A Q-factor much smaller than 1 means strong damping, *R* is dominant, so there is no ringing and the probe input largely behaves as the simple RC low-pass filter we discussed before. A high Q-factor on the other hand indicates very little damping, so an incident edge will excite strong, long lasting resonant ringing in the probe (at the frequency f_{res} given above). For the case just in between (close to a Q-factor around 1, also called "critically damped"), where the damping is too strong for real resonance, the RCL filter rise time is given by

$$T_{10/90} \approx 3.4 \times \sqrt{L \times C}. \tag{31}$$

In any case we see that reducing the parasitic inductance is always in our interest, since there is little we can do about the resistance of the system under test[45].

We have several possibilities to minimize the loop inductance: First, make the signal path as well as the ground connection as short as possible. A long cable that conveniently reaches a far-away ground point is an absolute no-no if we want anything close to decent performance. If we do have to use a short cable for the ground return, press it as close to the probe enclosure as possible, to avoid any unnecessary loop area. We may not always have much choice in the ground path inside the system under test, but choosing the closest possible ground attachment is usually a good idea. More than one ground connection is even better since it means we have several inductances in parallel, which further reduces the total inductance. Finally, as should be abundantly clear by now, having no ground connection at all is a capital crime in the world of probing, even though one may have heard people argue that "It worked just fine when I tried": What happens is that the return current will find *some* way back, probably through a huge detour comprising the fixture power supply and the scope power supply among other nasty

[45] At first glance increasing the probe capacitance to reduce the Q-factor and avoid ringing may look tempting, but all it will really do is trade in this resonance for severe rise time degradation because of the increased low-pass RC filter time constant. Reducing the inductance is our only valid option.

elements. The huge ground loop created by this will prevent any meaningful measurement above maybe a MHz or so (which probably was the speed those people were running their system at).

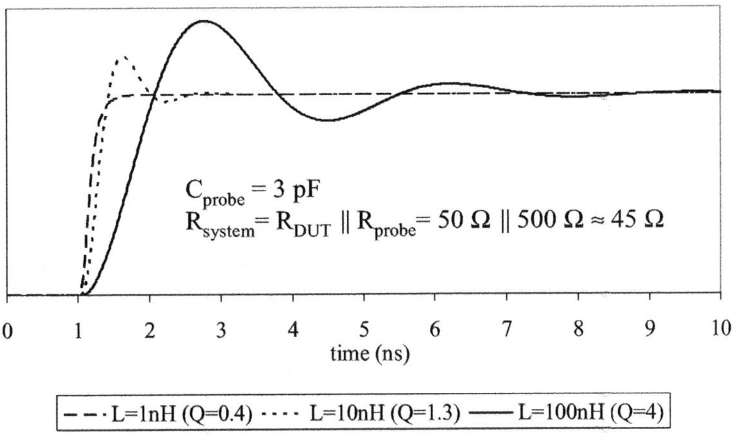

Figure 35: Acquired waveforms for different probe grounding schemes (resulting in different probe inductances), for a passive 500 Ω probe attached to a 50 Ω driver.

To get a *rough* estimate for the parasitic inductance of a given ground return loop we can use the following approximation formula:

$$L \approx 6 \cdot 10^{-7} \times D \times \left(\ln \frac{8 \times D}{d} - 2 \right), \tag{32}$$

where L (in H) is the inductance, D (in m) is the diameter of the loop (assumed to be a circle), and d (in m) is the diameter of the wire. Since the logarithm is a very slowly changing function, the inductance is mostly proportional to the loop diameter, the wire diameter being only of secondary importance. For example, if we have a loop of 1 cm diameter of a wire of 0.5 mm thickness, then the resulting inductance is

$$L \approx 6 \cdot 10^{-7} \times 10^{-2} \times \left(\ln \frac{8 \times 10^{-2}}{5 \cdot 10^{-4}} - 2 \right) \approx 18\,\text{nH}.$$

This is a huge inductance indeed for high speed purposes and will severely limit our measurement bandwidth. For example if we have a probe capacitance of only 1 pF (which is a pretty good and probably expensive probe), then the ringing frequency would be a measly

$$f_{res} \approx \frac{1}{2 \times \pi \times \sqrt{18 \cdot 10^{-8} \times 1 \cdot 10^{-12}}} \approx 380\ \text{MHz},$$

in other words, we would not be able to accurately measure anything even close to 400 MHz (or corresponding rise times of 1 ns or less).

To illustrate the preceding paragraph graphically, Figure 35 shows the acquired waveforms with different grounding schemes (long cable, short cable, minimum length ground pin).

4.4.5 Noise Pickup

In addition to degrading our probe rise time and causing ringing, the ground return loop has one more unpleasant effect: It picks up noise from electromagnetic fields, as shown in Figure 36. The process is simple magnetic induction – the signal loop acting as a coil that is traversed by a magnetic field changing in time. The induced voltage is proportional to the enclosed loop area, so here is another reason why reducing this area makes a lot of sense!

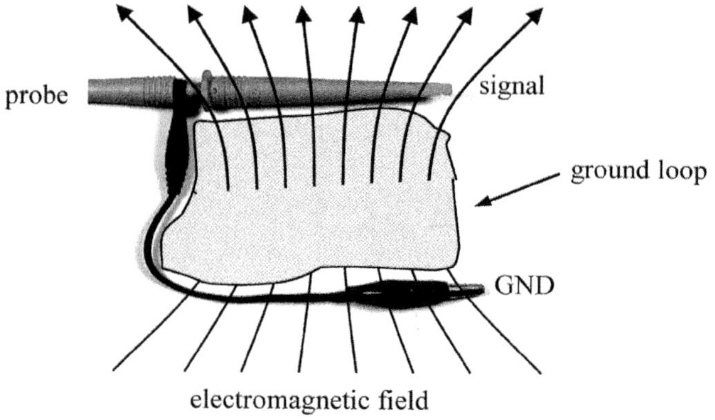

Figure 36: Noise pickup through the probe ground loop.

How can we find out if a strange-looking feature or some excessive noise on our waveform is caused by noise pickup? As a first step, disconnect the probe from the system under test and connect the probe tip directly to the ground lead. Without noise pickup we should not see any signal on the scope. Any signal we do see is external noise coupling into our loop. To further prove that this is the source, change the size of the loop (e.g. by squeezing the ground return closer to the probe tip) – the noise amplitude should change accordingly.

Since only magnetic fields *traversing* the loop can cause an induced signal, rotating the loop so the field lines are simply passing by is another way to reduce the size of the induced noise. This directional sensitivity can also help in tracking down the source of the interference.

4.4.6 Avoiding Pickup from Probe Shield Currents

Figure 37 shows another practical case of external noise coupling into the oscilloscope (or other measurement device). Chances are every test and characterization engineer has been bitten by this situation at least once. The troublemaker here is the huge loop that is closed by the separate ground connections of the system under test and the oscilloscope. If there is even just a weak field present (e.g. 60 Hz noise from other power lines or from fluorescent lights) it will induce a sizeable loop current I_{noise}. This in turn will produce voltage noise because even the best cable connection has some ohmic resistance. The important parameter here is the resistance R_{shield} of the signal return path (the outer shield for a coaxial cable); it will cause a voltage offset V_{noise} between the system ground and the instrument ground:

$$V_{noise} = I_{noise} \times R_{shield} \tag{33}$$

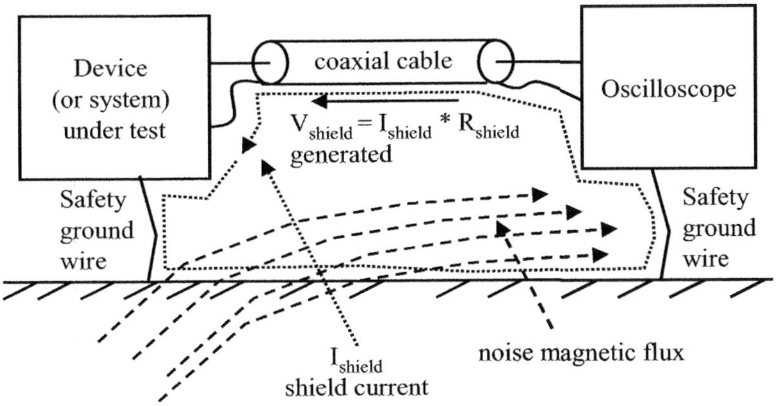

Figure 37: Noise caused by probe shield currents.

One can try several things to mitigate this noise:

- Reduce the current loop by attaching scope and system to same power outlet and keep the power cables close to each other. This is good practice anyway.

- Add a shunt capacitance between the scope and logic ground on the test fixture. This causes more of the noise current to flow through the shunt and less through the shield.

- Put a big inductance in series with the shield. This raises the inductance of the probe shield, lowering the current. An easy way to do this is to wind the probe cable a few times through a ferrite choke. This method is useful between maybe 100 kHz and 10 MHz – below that range the achievable inductance is too small, above the ferrite material ceases to be effective.

- If using a passive 10:1 probe, replace it with a 1:1 probe. A 10:1 probe attenuates actual logic signals and makes shield voltages appear relatively ten times larger. But 1:1 probe often has much larger parasitics, and the circuit may not like the increased load.

- At the very last, the best method is to use true differential probing. A differential probe has both of its connections floating (i.e. not connected to ground or the shield), thus it picks up the correct ground level at the fixture. This is one of the major advantages of differential probes compared to single-ended probes.

4.4.7 Rise Time and Bandwidth

As we have just seen, parasitics in the probe front-end section limit the maximum bandwidth of the probe and thus increase the rise time of the signal. If our probe is an active probe then the amplifier will compound this effect, and in any case the oscilloscope will degrade the signal further. As usual, those rise times add up approximately as root-mean-square, that is

$$T_{r,display} \approx \sqrt{T_{r,signal}^2 + T_{r,probe}^2 + T_{r,scope}^2}\,, \tag{34}$$

where $T_{r,display}$, $T_{r,signal}$, $T_{r,probe}$, $T_{r,scope}$ are the displayed rise time, the "real" signal rise time, the probe rise time, and the scope rise time, respectively.

The equivalent bandwidth of the scope-and-probe combination[46] is of course[47]

$$BW_{scope+probe} = \frac{1}{\sqrt{\frac{1}{BW_{probe}^2} + \frac{1}{BW_{scope}^2}}} \tag{35}$$

Note that for high-performance probes the specified probe bandwidth only holds true with optimum grounding (minimum length ground lead). It should be clear that between scope and probe the weaker link determines overall performance – a low-bandwidth probe (or even a good probe, but with long ground connection that causes ringing and bandwidth degradation) can obliterate the performance of even the best – and most expensive! – oscilloscope. Since probes are typically much less expensive than a scope of similar bandwidth it always makes sense to get a probe with somewhat higher bandwidth rating than the scope.

4.5 Differential Signals

So far we have always regarded our signals as being single-ended (i.e. a single signal line referenced to ground). However, with faster data rates differential signaling is becoming more and more prevalent and is most likely to replace single-ended signaling in most applications. On the other hand, many traditional oscilloscopes and probes are only single-ended. Even though we haven't gotten into jitter – we will do so in the next chapter – it probably seems very plausible even now that in order to properly characterize all the jitter characteristics of a differential signal, the acquisition must be done differentially as well.

4.5.1 Probing Differential Signals

The most straightforward method for this is of course to use a true differential probe, which virtually always is an active probe. There is no

[46] We should note that some scope manufacturers directly specify on their probes the total bandwidth of scope plus probe, assuming of course that one uses one of their probes with the intended scope of theirs. For high-bandwidth oscilloscopes this actually makes a lot of sense because in this performance range scope and probe are usually optimized for each other (e.g. for minimum ringing), so using a third-party probe can often result in sub-optimal performance.

[47] One gets this result by simply replacing the rise times with the bandwidth in the formula for adding bandwidth, using $T_r \approx k / BW$, and assuming that the k factors are the same for probe and for scope (which may not be exactly the case).

fundamental design difference between a single-ended active probe and a differential one apart from the fact that the input of the latter is freely floating and referenced to the second input, while the former has ground as its reference. Both probe types transmit the same type of signal to the oscilloscope, i.e. the scope does not care if a differential or a single-ended probe is attached to it. So in other words, any active differential probe, while giving the flexibility to probe differential signals, can also be employed to probe single-ended signals without restriction.

It also means the same limitations apply, namely bandwidth restrictions – even the most advanced commercial differential probes max out around 12 GHz and are thus ill suited for data rates beyond 6 Gb/s. And since they employ active elements (the amplifier), they inevitable add additional noise and jitter onto the signal. Still, as long as these limitations are beyond the required performance, these probes are the tool of choice for differential probing. They also offer a high common mode rejection ratio (CMRR) at lower frequencies (a few MHz), but the CMRR falls of rapidly with increasing common signal frequency.

The great advantage of direct measurement of the differential signal (using a true differential active probe) is that all timing and jitter measurements are then absolutely identical to single-ended measurements – after all, what the probe delivers is a single-ended signal proportional to the difference between the two input signals. Second, with a high impedance probe we can directly probe signals in a system, e.g. signals running on a bus, without adding excessive load (which otherwise may make the system fail since additional loading causes additional attenuation).

Unfortunately where we encounter differential signals is predominantly at highest speeds (where the advantages of differential signaling are becoming absolutely necessary to retain enough noise margin), while at the same time those probes – being active devices – are restricted in bandwidth. In other words, just where we would need them most, active differential probes hit their performance limit.

Most high-end bit error rate testers also offer differential capabilities (both for drive and for receive). Compared to oscilloscope probes their task is made somewhat easier by the fact that they do not intend to reconstruct the full waveform, but only need to trigger at a certain threshold.

Another option is to feed the two single-ended components (true and complement signal) of the differential signal into two separate channels on the oscilloscope (let's call them channels A and B), and use the built-in math functions to display the difference signal $D = A - B$. That way one is not limited by the maximum active probe performance (12 GHz at best), but gets the full bandwidth of the oscilloscope – especially in the case of equivalent-time sampling scopes a large benefit. Of course this will put a 50 Ω load on the circuit, which is usually okay in a device characterization situation (where the only load is the scope, i.e. no in-circuit probing is necessary), but would otherwise be unacceptable. It also only works well if the signals don't carry excessive common mode components (depending whether they are DC balanced or not, AC-coupling could potentially remove this problem). Otherwise this is a perfectly viable option as long as we consider two things:

In order for this scheme to work the two channels (probes, cables, and scope input amplifiers and samplers) have to be extremely well matched, both in delay as well as in gain. Otherwise a sizeable portion of spurious common mode signal will show up (that is not present on the real differential signal), in other words, the common mode rejection ratio will be very poor. Out of this reason until a few years ago single-ended probing of differential signals was usually strongly discouraged, since especially analog scopes had no way of tightly controlling their amplifiers' gain and gain linearity. But today, where digital high-end instruments can employ sophisticated gain and linearity compensation and calibration, gain mismatch has largely become a non-issue. As for timing, given a good cable vendor it is possible to obtain cable delay matching down to 1 ps, which is enough even for the fastest speed we encounter today (delay should be matched to within a small fraction of a rise time, and even 12 Gb/s data signals have 20/80 rise times around 30 to 40 ps).

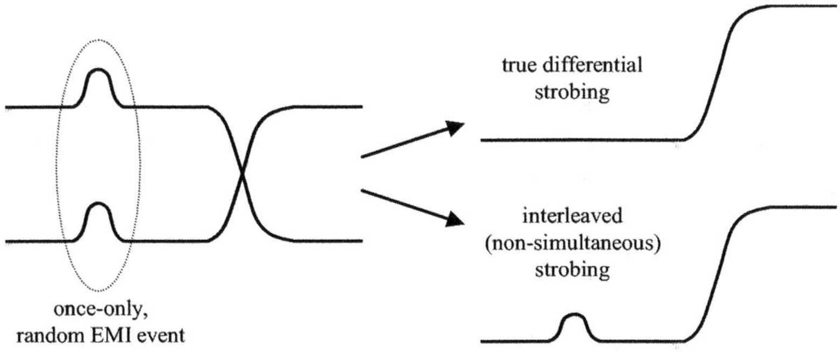

Figure 38: For differential signals, true differential (simultaneous) strobing of the signal pair can yield very different results to interleaved strobing.

Second, in order to look at random jitter and noise, the scope must be able to strobe both channels simultaneously. Otherwise the math waveform will not display the true differential signal because each point consists of two data points (from *A* and from *B*, respectively) taken at different times – so the instantaneous random noise would be completely unrelated between those two. As an example, look at Figure 38. The signal is experiencing a common mode spike that moves the levels of both single-ended signals simultaneously and in the same directions. If both *A* and *B* are sampled at this same instant, the scope will correctly display the differential signal, unaffected by the spike (because it cancels in the difference). But is A is sampled at this instance, and B is sampled at the next pattern repeat (assuming this time no spike occurs), then the spike is only present at A and thus on the display – in contrast to the real situation on the signal.

As a matter of fact, many oscilloscopes employ interleaved (non-simultaneous) strobing, either to save cost, or because of hardware limitations. Unless we know for sure what our situation is, how can we check if scope does simultaneous strobing or not? The easiest possibility is to take a differential pattern generator and let it drive a pseudo-random bit stream (many generators have this capability built in; otherwise they may allow to program an arbitrary pattern). Feed one channel (true) into scope channel A, the other (complement) into scope channel B. Trigger the scope on a clock running at the data rate speed (i.e. don't use a trigger that only occurs once per pattern!). Set the scope up to display $D = A - B$. If sampling is simultaneous, D will display as a two-level eye diagram[48] as in Figure 39(a) because the signals at the same instant are always the opposite of each other: Either $A = low$ and $B = high$, thus $D = high$; or $A = high$ and $B = low$, thus $D = low$). If not, then *three* levels will appear because the level on A (taken at some time) and the one on B (taken at some other time) are completely unrelated to each other (third level: $A = low$ and $B = low$, or $A = high$ and $B = high$, both resulting in $D = midlevel$ (zero) between high and low), as shown in Figure 39(b).

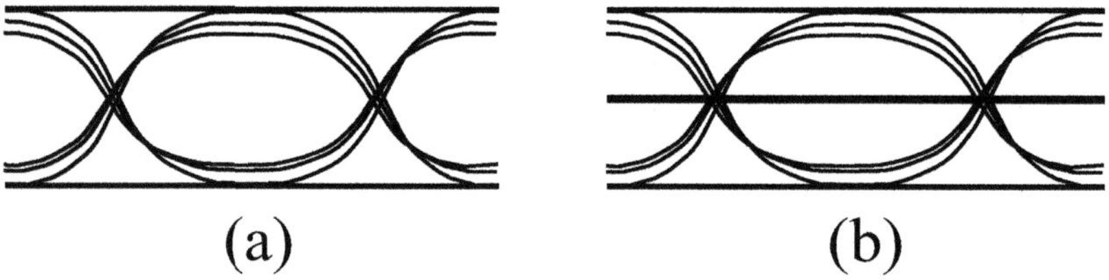

Figure 39: An eye diagram of the differential signal, acquired through two single-ended channels, reveals if the oscilloscope acquires the two components simultaneously (a) or interleaved (b).

A final word of caution: Some scopes on the market do simultaneous strobing when the deskew adjustment between their channels is set to zero, but when this adjustment is non-zero (even a single ps is enough!) they switch over to interleaved strobing. In this case if we need to remove skew between those two channels our only option is to add an adjustable hardware delay line (see section 5.2.7) in front of the scope inputs and leave the internal scope deskew setting at zero.

[48] We will talk about pseudo-random patterns and about eye diagrams in the next chapter.

Second, even when an oscilloscope acquires both channels simultaneously this does not necessarily mean the strobe jitter of those channels is correlated, because some part of the two strobe delivery paths may be separate. In other words, the observed random jitter of the displayed differential signal may or may not be the true differential jitter on the actual signal.

4.5.2 Single-Ended Measurements on Differential Signals

The remaining problem is how to relate the jitter measured on the difference trace on the scope to the actual jitter on the real differential signal. For deterministic, static jitter types like data dependent jitter or duty cycle distortion this is trivial – they are the same.[49] Unfortunately this is not necessarily true in the case of random jitter or any type of jitter (uncorrelated periodic jitter comes to mind) that requires single-shot acquisition:

Let's assume our single-ended signals are S_1 and S_2, and we let the scope (or tester) calculate the differential signal M as

$$M = S_2 - S_1 . \tag{36}$$

If S_1 and S_2 are sampled simultaneously then M corresponds to the real differential signal and we are done – just perform any measurement and analysis on M instead of the single-ended signals. But many scopes – and equivalent-time sampling scopes in particular – interleave the sampling between the two channels; e.g. the scope may first acquire a full trace of S_1, and then a full trace of S_2, or maybe acquire one sample of S_1 and then one sample of S_2, and so on. Nobody can guarantee that the instantaneous jitter at the first sample instant is the same as the jitter at the second instant – especially for random jitter. In fact, we can almost guarantee it is not!

This is less of a concern for data dependent jitter or duty cycle distortion because it will not change from one run (or sample) to the next – per definition it only depends on the pattern driven. Thus we can average the two curves over several acquisitions (to get rid of uncorrelated and random jitter), and the differences of the averages will be identical to the average of the differences. Done.

There is no such simple way out for random jitter and other uncorrelated jitter. Let's have a look on Figure 40 which shows the two single-ended components of the differential signal:

In Figure 40(a) S_1 and S_2 jitter together (their timing jitter is correlated). This is a common case when the signals are generated by a truly differential,

[49] We are getting a little ahead of ourselves here. For more details about the different types of jitter refer to the next chapter.

low-noise driver that does not add much jitter (through noise) on its own, so the timing jitter mostly comes from earlier stages of the signal generation and will affect both signals' timings the same way. In this case the timing jitter of the true differential signal is equal to the timing jitter of each of the single-ended signals, which is easily measured. This case also applies to all types of data dependent effects as long as the two signal paths are comparable (same bandwidth etc.).

Figure 40(b) displays the opposite case – here the jitter of S_1 is the opposite of the jitter on S_2 (i.e. when one signal is early, the other one is late). The most frequent cause of this situation is common noise on both signals (e.g. from crosstalk or EMI) that makes both signals move higher (or lower) in lockstep, as indicated in Figure 40(b). In this extreme the differential signal has no timing jitter whatsoever (the crossing point moves up and down in voltage, but stays still timing-wise), while each single-ended signal on its own certainly has. Measurement of the single-ended jitter here would be completely misleading since it does show significant jitter.

Finally, there is Figure 40(c) where both signals' jitter is completely independent from each other. This can come e.g. from random noise on the final driver stage of the signal source. Let's assume at some given transition signal S_1 has the ideal timing, while S_2 is late by some amount – as shown in Figure 40(c). The differential crossing point is moved in both voltage and timing, but the timing movement is only *half* of the displacement of S_2. Thus the differential jitter of M contributed by S_2 is only half the single-ended jitter on S_2. At the same time, S_1 also contributes some jitter. If we assume the jitter on S_1 and S_2 to be the same amount (very likely if the two paths are designed the same), and further assume the jitter to be purely random (Gaussian distribution), then their values will add up as RMS, i.e.

$$jitter_M = \sqrt{\left(\frac{jitter_{S1}}{2}\right)^2 + \left(\frac{jitter_{S2}}{2}\right)^2} = \frac{jitter_{S1}}{\sqrt{2}} . \qquad (37)$$

In other words, the jitter on the true differential signal is smaller than the measured single-ended jitter by $\sqrt{2}$.

So with the three cases above for given (measured) single-ended jitter, the true differential jitter on the differential signal can either be zero, or equal to the single-ended jitter, or smaller (by $\sqrt{2}$) than the single-ended jitter. In any practical situation there will always be some mixture of those three cases (although one can be dominant), but in all of the cases the true differential jitter is always *smaller* or at worst equal to the measured single-ended jitter. This means that if we only measure single-ended, the result is a worst-case, pessimistic estimate of the true differential jitter.[50]

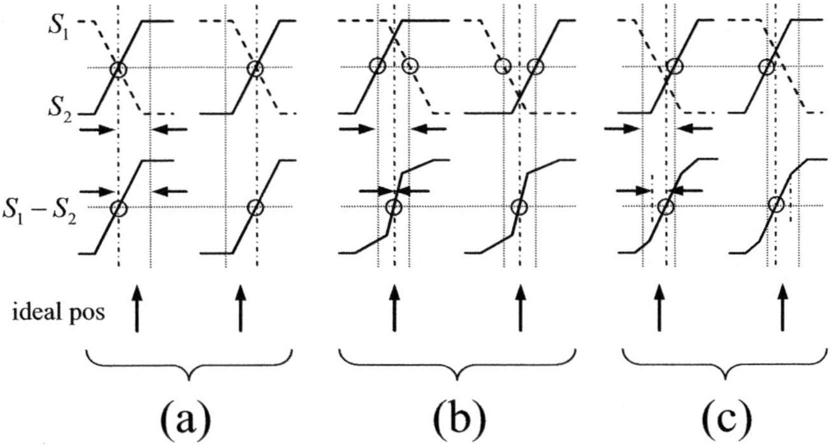

Figure 40: Possible relationship between the random jitter of a differential signal vs. the random jitter of its single-ended components: (a) correlated, i.e. differential jitter equal to single ended jitter, (b) anticorrelated, thus differential jitter zero, (c) uncorellated, thus differential jitter smaller than single ended jitter.

4.5.3 Passive Differential Probing

If we absolutely need to determine the uncorrelated differential jitter (random, periodic, uncorrelated deterministic) and only have single-ended measurement capability, one way is to use a *passive* Balun (see section 5.2.2) to convert the differential signal into a single-ended ones. Most Baluns will restrict our bandwidth and will distort the signal to some extent, but we are not out to measure data dependent effects with that setup anyway. Rather we can take advantage that – being a passive element – other than limited in bandwidth, random jitter is not influenced by such a device, and not worry too much about other signal distortions. Data dependent effects can later be measured single-ended without the Balun in the path.

[50] This favorable relationship between singled ended and differential jitter is also one of the reasons (but by far not the only one!) why differential signaling is so popular for higher data rates or noisy transmission situations because it provides a reduction in jitter compared to the single-ended case, of course at the cost of doubling the channel count.

5. ACCESSORIES

5.1 Cables and Connectors

There may be no other part of the signal chain that is so often overlooked as the humble coaxial cable. As long as speeds are moderate, it is merely seen as a passive interconnect that transports the electrical signal from the source (in our case the device under test) to the receiver (the oscilloscope or similar instrument). But as a matter of fact, the electrical quality of the cable used determines to a significant amount the high-frequency performance of the overall measurement system.

Coaxial cables – having a homogeneous geometry and therefore homogeneous capacitance and inductance per unit length – act as transmission lines, in our case most likely with a characteristic impedance of 50 Ω. As any real transmission line they are subject to losses – DC (ohmic) resistance, skin effect, and dielectric loss. Those losses give rise to an attenuation of the signal, which is usually specified in dB per length (e.g. dB per meter), i.e. on a logarithmic scale. One of the advantages of this scale is that the loss numbers for series of cables simply add up, i.e. when cable 1 has a loss of 2 dB and cable 2 has a loss of 3 dB at a given frequency, then the series combination of those two cables will have a loss of $2+3=5$ dB at this frequency. As we recall, a loss of 3 dB corresponds to a 30% reduction in amplitude, and 6 dB mean we lost already half of our signal. In the previous chapter we saw that skin effect losses increase with the square root of frequency and dielectric loss linearly proportional to the frequency, so at higher speeds those effects will become more and more important. High-quality cables have usually specified their total losses at several different frequencies.

5.1.1 Cable Rise Time/Bandwidth

For digital signals in particular there is more to losses than a simple reduction in amplitude. Keeping in mind that frequency is the inverse of amplitude, it is immediately plausible that losses affect the strongest the part of a waveform at or immediately after a transition because short time scales correspond to high frequencies. The outcome is that those frequency dependent losses degrade (increase) the rise time of edges, which in turn not only affects the accuracy of rise time measurements, but it worsens data dependent timing errors (see later) and closes the data eye because the signal does no longer reach its full level before the subsequent transition: Since at least in theory both skin effect and dielectric loss only get smaller, but never really disappear with decreasing frequency (increasing time scale), they are effective even long time after some transition; in reality the exponential behavior of dielectric effects makes them negligible rather soon, while the square root behavior of the skin effect causes it to stick around for much longer.

It should be obvious that a longer cable of the same type as a shorter one means more losses, but what is the exact dependency?[51] Let's say we have a cable of some length that – due to skin effect and dielectric loss – has some rise time T_r. What happens if we put two of those cables in series, e.g. to be able to connect our system under test to a bulky oscilloscope that we can't (or don't want to) approach too close to the system? We may remember the RMS rule (adding up the squares) to add up rise times from the previous chapter. It is tempting to apply this rule blindly to the new situation, but this neglects the fact that the rule is only valid for Gaussian (and closely valid for simple one-pole exponential) edges – unfortunately neither skin effect nor dielectric loss produce Gaussian edge shapes.

For dielectric loss it turns out that the rise time instead increases *linearly proportional* to the length of the cable. In other words, if we double the length of the cable, the equivalent rise time of the (longer) cable will be twice as large as before. This is a far stronger increase than the $\sqrt{2}$ increase that we would have expected from the RMS rule!

But it gets even worse for skin effect – the rise time in this case increases *proportional to the square* of the cable length. This odd behavior becomes more plausible if we remember that in frequency domain the skin effect increases with the square root of the frequency, thus in time domain it disappears proportionally to $1/\sqrt{time}$. Skin effect is based on the ohmic resistance of the cable material (aggravated by the inhomogeneous current distribution), and ohmic resistance doubles when the cable length doubles. Thus in order for the skin effect of a cable of a certain length to reduce to the same value as a cable of only half that length, we have to wait *4 times* as long, or in other words, the skin effect loss increases with the square of the length.

The strong dependency on the cable length for both skin effect and dielectric loss illustrates well the paramount need for (and the great benefits associated with) keeping cables as short as possible. For high-end oscilloscopes there are often extenders available that allow us to bring the (small) sampling head as close as possible to the system under test – the

[51] As stated before, in the frequency domain (i.e. corresponding to RF type signals) things are easy: If the loss for some cable for some frequency is x dB, then a cable twice as long will have 2x dB loss. But digital signals, which contain a very wide band of frequencies (each of them experiencing a different amount of loss), the situation becomes more complicated to describe, especially in the time domain.

extender cables carry only the strobe signals whose exact shape is only of secondary importance, but it enables us to minimize the analog signal path from the system through the cable to the sampling head.

Given the rise time of the cable used (or maybe even the relative contributions of skin effect and dielectric loss), the output rise time for a given input signal rise time can still, at least approximately, be calculated using the RMS rule, as long as we calculate the cable rise time based on the behavior above (linear and quadratic dependency, respectively):

$$T_{r,out} \approx \sqrt{T_{r,in}^2 + T_{r,cable}^2} \approx \sqrt{T_{r,in}^2 + T_{r,skin}^2 + T_{r,dielectric}^2}\,. \tag{38}$$

The approximate bandwidth is then given as usual by

$$BW_{cable} \approx \frac{0.33}{T_{r,cable}}\,. \tag{39}$$

5.1.2 Skin Effect Compensation

By now it should be obvious that the main enemy of digital signal transmission through a cable is less the absolute attenuation than rather the dependency on the frequency, because this is what distorts the signal edges, and what causes data dependent jitter.[52] A possible workaround is to make a tradeoff between static losses (ohmic, DC) and variations due to skin effect. Figure 41 shows a cross-section of such a skin-effect compensated cable. Its center conductor (which has a much smaller surface than the shield and thus contributes most of the skin effect losses) consists of a rather resistive core, plated with a thin layer of smooth and highly conductive material (e.g. silver). The high core resistance means that even at low frequencies (where the current takes the path of least ohmic resistance) most of the current will already flow in the thin outer layer, so when skin effect takes hold at higher frequencies, there is only little change in the current distribution, and therefore the effective resistance does not change much. Of course this is only true as long as the skin depth is larger than the thickness of the outer layer – once it becomes smaller, the same old $\sqrt{frequency}$ behavior takes hold again. Overall, this provides good compensation of skin effect (i.e. a very flat loss profile vs. frequency) up to a certain maximum frequency.

[52] Constant (frequency independent) losses merely reduce the signal levels at each instant by a constant factor, but they do not cause any waveform distortion. Such losses can easily be taken into account by adjusting the driver levels and/or the receiver threshold levels.

One can improve this scheme further by using some ferromagnetic material for the core. That's one reason why cables with silver-plated steel cores are so popular (apart from their mechanical strength). Looking at the skin effect formula for a cylindrical cable in section 1.6.5.2 we see that the skin depth is inversely proportional to the square root of the magnetic permeability μ, which for steel can be thousands of times higher than for non-ferromagnetic metals. As a result the current gets pushed out of the steel core already at very low frequencies, flowing almost entirely in the copper (or silver) plating, and the frequency response is flat from low frequencies on up to where the skin depth becomes smaller than the thickness of the plating.

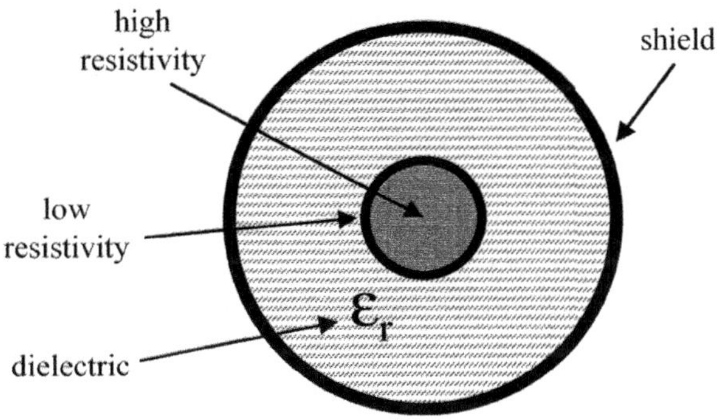

Figure 41: Cross-section of a coaxial cable with skin-effect compensation.

5.1.3 Dielectric Loss Minimization

For dielectric loss the most straightforward approach is to use a lower-loss dielectric. Vacuum (or air) would be ideal, but except for absolutely rigid transmission line structures it would pose insurmountable technical challenges. So the next best thing, which is employed in high performance cables, is to use foamed Teflon. Teflon has very small losses up to high frequencies to begin with, and they are further reduced by the amount of air contained in it because the signal now sees a mixture of Teflon (low losses) and air (close to zero losses). Of course the size of the air bubbles in the Teflon must be orders of magnitude below the wavelength of the highest frequency component to be transmitted so the dielectric really acts as a homogeneous body.

The only downside of this foamed dielectric (as opposed to a solid one) is reduced mechanical ruggedness. High-performance cables don't like to be flexed too often and will wear out rather fast. This wear-out can manifest itself in changes in delay, loss, and signal shape when the cable is flexed.

For the coaxial connectors, dielectric loss minimization means replacing the solid dielectric with minimum-size standoffs that are just sufficient to keep the center conductor in place. The price is again reduced mechanical stability.

5.1.4 Cable Delay

As long as we do not have to deal with propagation delay matching between the two lines of a differential pair, the absolute delay of our cables is normally of secondary importance. What we do care however are changes in the delay during the measurement.

All dielectrics exhibit more or less pronounced changes of their dielectric constants with temperature – and thus changes of the propagation delay since the propagation speed varies with $\sqrt{\varepsilon}$. So keeping the ambient temperature constant during a long measurement becomes very important for highest accuracy requirements.

Second, all cables show some change in delay when they are bent and flexed since the dielectric deforms and thus the cable geometry changes. For quality cables the maximum amount of change (for flexure down to the minimum allowed bend radius) is usually given in the cable specifications. For good cables the change is in the order of 1 ps, but cheaper, lower-performance cables can produce many times that amount. In any case for highest accuracy it is advisable to avoid any movement of the cables during measurement; another option is the use of a rigid or semirigid cable assembly (of course at the cost of setup flexibility and ease of use).

5.1.5 Connectors

Just as important as it is to get the signal *through* the cable (or some other component of the path) it is to get the signal *into* and *out of* the cable. This is where connectors come in. The best, highest bandwidth cable will not help us if we cannot connect it to the rest of the setup with equally high bandwidth.

For high-speed, high-accuracy laboratory environments – where performance is paramount and the number of connections is usually small – there exists a relatively small number of standard connector types, each geared towards a specific frequency range and level of ease of use.[53] Table 1 shows an overview for the common types (ordered from lowest to highest

[53] The situation would be different e.g. for consumer applications or large-scale production setups where cost per pin/channel has to be low and often a large number of connections has to be made.

performance): BNC, Type-N, SMA, 3.5 mm, 2.92 mm, 2.4 mm, 1.85 mm, and 1.0 mm.

Some common trends are visible when going to higher and higher performance connectors (and the general trends hold true for connectors other than the ones discussed as well):

First, all the connectors are coaxial, impedance-controlled designs. Good impedance control is easiest achieved for straight barrels, so any bends (i.e. connectors that have a 90 degree launch into the cable) risk to introduce discontinuities unless the manufacturer went to great lengths in the geometric design to keep the impedance constant throughout the path within the connector. Thus, when in doubt it is always better to use a connector with a straight launch into the cable.

Second, like any transmission line, connectors exhibit losses that limit high-frequency performance. To reduce the dielectric loss contribution, good connectors must either use a high-performance dielectric like Teflon, or try to minimize the amount of dielectric altogether. Thus in the highest bandwidth types one will only find small standoffs that keep the center pin in place, rather than a full solid barrel of dielectric. This is also a giveaway when looking at an unknown connector – e.g. SMA and 2.92 mm are mechanically compatible (same thread, can mate to each other) but the (higher-bandwidth) 2.92 mm connector will contain largely air.

Third, in order to avoid any dispersion (and thus signal edge distortion) in the connector, the diameter of the connectors must be much smaller than the wavelength of the highest frequency component to be transmitted (it is the inner diameter of the outer (shield) conductor that counts). Otherwise there will be "strange" modes of propagation – cavity resonances – that introduce loss and distortion. So it is not to surprising that BNC connectors have a much larger diameter than SMAs, and those have larger diameters than e.g. 2.4 mm connectors (in fact the mm number gives this diameter; it does not tell anything about the outer thread diameter, so e.g. 3.5 mm and 2.92 mm connector have compatible threads.

Finally, any slots or holes in a connector cause resonances and radiation losses – thus only the low-bandwidth BNC type has such slots in the outer conductor, and the highest-bandwidth connectors avoid even slots in the center barrel of the female type that mates with the male plug. This of course requires extremely high geometrical precision during manufacturing to assure reliable contact without the compliance provided by a slotted and thus elastic receptacle.

Table 1: Commonly used coaxial connector types.

Name	Bandwidth GHz	Remarks	Picture
BNC	2...4	Cheap, easy to connect (latch-on). Slots leak radiation. 50 Ω and 75 Ω versions.	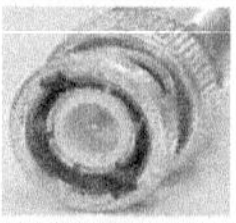
Type-N	up to 18	Often used in RF equipment (e.g. sine generators); 50 Ω and 75 Ω versions.	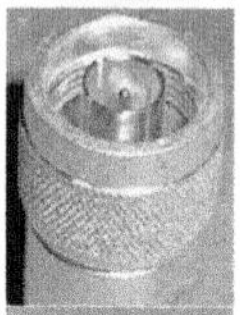
SMA	12...20	Widely used; poorly standardized tolerances.	
Super-SMA	26	Improved, high-precision SMA version.	
3.5 mm	34	Mechanically compatible to SMA.	
2.92 mm (K[54])	40	Mechanically compatible to SMA.	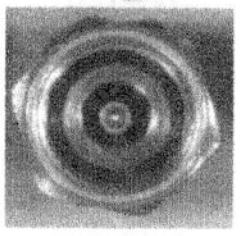
2.4 mm	50		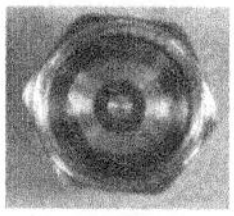
1.85 mm (V[54])	65	Mechanically compatible to 2.4 mm	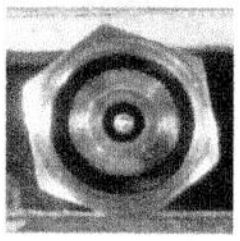
1 mm	110		

[54] "K" and "V" are Anritsu's copyrighted designations for their 2.92 mm and 1.85 mm connectors, respectively.

A final word about good care for connectors (and any other accessories of our setup): High-bandwidth connectors are very sensitive to abuse and their performance will degrade quickly if not handled properly – and those changes may go unnoticed for a long time while degrading our signal integrity and our measurement results. Please keep the rubber dust caps they came with on whenever the connector is not in use. Use a torque wrench of the proper torque to tighten them – this assures a solid connection (and good signal integrity) while at the same time preventing damage from over-torqueing. When tightening the screw, always turn the outer barrel and *not* the connector (or cable) itself, because that would grind down the internal mating surfaces. To clean a dirty connector use some alcohol on a cotton swab, but don't touch the center pin of male 3.5 mm, 2.4 mm or faster connectors – there is not much dielectric to hold them in place and we would risk breaking or dislocating them.

5.2 Signal Conditioning

5.2.1 Splitting and Combining Signals

Sometimes we need to send the same signal into more than one input. A typical example is an equivalent-time sampling scope (that needs an external trigger) when we want to trigger on the signal itself. Or we may want to monitor a signal on a bus by picking off a small portion of it.

The worst thing we can do in this case is to use some of the (readily available) T-connector pieces (shown in Figure 42(a)). As a 1:2 splitter they are horrible because they make no effort at all to match input and output impedances: The incoming signal hits a combination of two transmission lines in parallel, so strong reflections will occur. Two better solutions are the power splitter (Figure 42(b)) and the power divider (Figure 42(c))[55]. The resistors in series combination with the transmission line impedances provide the necessary impedance matching. While the power splitter only looks like 50 Ω for the input port (left side in Figure 42(b)), the power divider is matched for any of the three ports.

[55] We should mention, however, that the impedance matching comes at a price, namely ohmic power dissipation in the resistors. The output amplitude on each output is only half the input, so overall half the power is lost. The T-connector on the other hand does not dissipate any energy at all (it only reflects and transmits it), and the transmitted power is thus higher than for the other two solutions (transmission coefficient of 66% vs. 50%). If the signal sources provide matched source termination, thus swallowing the reflections, it can sometimes be the solution of choice to maximize the transmitted signal amplitude.

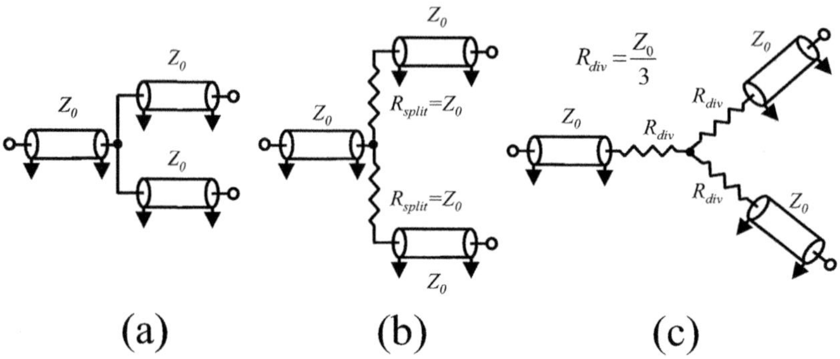

Figure 42: Different designs for broadband signal splitters: (a) simple T-element, (b) power splitter, (c) power divider.

Of course, when driven in the opposite direction (two signals driven in, one coming out), a divider will act as a power combiner, i.e. it will perform a summing operation on its two input signals. This can be useful e.g. to look at the common mode voltage of a differential signal if the oscilloscope can't add up two channels, or for adding some controlled amount of noise to a signal (see also section 9.2.13).

Beware of any splitters (or dividers) that are not explicitly sold for time domain applications! Splitters for RF type applications are normally only designed to work over a narrow frequency band (the main goal here is usually minimum losses, so they do not employ lossy resistors but rather matched-length transmission line couplers and other "esoteric" stuff) and they won't perform at all when presented with a – wide-band! – digital data signal.

5.2.2 Conversion between Differential and Single-Ended

When dealing with differential signals there is often a need to either convert a single-ended signal into a differential one (e.g. we want to generate a differential signal but almost all RF generators are single-ended) or vice versa (e.g. we only have single-ended inputs on our oscilloscope but need to measure a differential signal). A Balun (from *bal*anced-*un*balanced[56]) is a useful accessory that achieves this conversion. The choice here is between active and passive elements.

The block diagram of a passive Balun is shown in Figure 43(a). While slower-speed passive Baluns (up to a GHz or so) actually use real wide-bandwidth transformers, faster designs make use of transmission line effects –

[56] Depending on the application (especially RF vs. digital), "differential" is also called "balanced", and "single-ended" is called "unbalanced".

an example[57] which works from a few kHz all way up into the multi-GHz region is shown in Figure 43(b). Because of their design, passive Baluns always have some lower frequency cutoff in addition to their high-frequency limit. Their main attraction comes from the fact that – being passive devices – they don't add any random jitter (but they may exhibit nonlinearities and data dependent effects).

An active Balun can be a simple differential high-speed op-amp circuit (including feedback of course). The advantage is that it can potentially operate all the way down to DC, but on the other hand any active element inevitably adds some random noise and nonlinear distortion to the signal (although those effects may be small enough to neglect them if using top-of-the-line components and good circuit design). In addition, while passive Baluns work in both directions (differential-to-single-ended conversion as well as single-ended-to-differential conversion), active Baluns typically can only perform either in one or in the other. This is not necessarily a disadvantage since it provides isolation against reflected signals.

Again, beware of components geared towards RF type applications (which have been around for a long time and are widely available) – most likely they will not work over the wide band of frequencies that a digital application needs. Always check the operating range and the gain (loss) flatness.

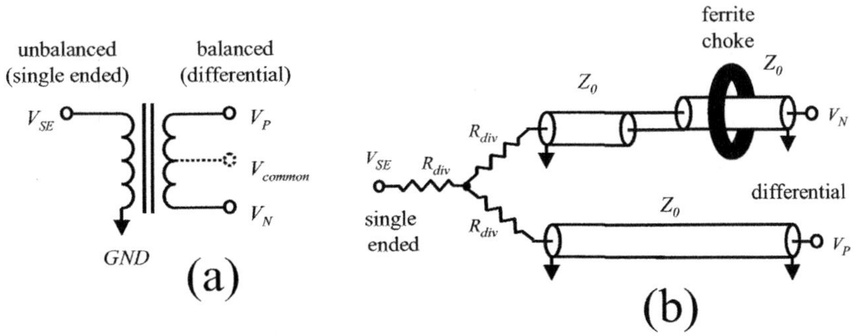

Figure 43: Different passive Balun designs: (a) transformer, (b) crossed transmission lines (semirigid coax) with RF choke (the second transmission line assures delay matching between true and complement path).

[57] Why the setup works may not be immediately obvious, so it deserves some explanation: The RF choke assures that the current coming out of the ground conductor on the balanced side *must* come back through the center conductor (and not through some other ground path) since every other path would loop around the choke and thus have huge inductance – which effectively prevents current through those paths down to rather low frequencies (a few kHz). For higher frequencies, where the ferrite loses its efficiency, the small parasitic inductance of the other ground paths takes over that role, since the shield in any case provides the lowest possible total path inductance..

5.2.3 Rise Time Filters

In digital (time-domain) applications, the only filter type used regularly is the low-pass filter[58]. Such a filter increases the rise time of the transmitted signal (in the frequency domain this corresponds to an attenuation of all frequencies above a certain threshold). Typical use cases are adjustment of the rise time of a signal source so the signal matches more closely the intended application, or simulation of the limited bandwidth of a slower receiver when measuring with a higher-bandwidth oscilloscope.

Some caveats apply here as well. First, depending on the actual design, such filters may be reflective or absorptive. The first type simply reflects the unwanted signal components – which can spell trouble of the driver cannot "digest" (terminate) strong reflections or if there are other impedance mismatches or parasitics in the signal path (because they will re-reflect those components to the receiver). The latter type absorbs them so they cannot interfere with our measurement. This is of course the preferred type, because in order to provide pure absorption (no reflections at the filter input) those filters have to provide matched 50 Ω impedance.

The second trap is the filter type. When one looks for filters (e.g. browsing the internet), chances are most of the search results will be filters for RF (frequency domain) applications, and again they are usually not well suited for the requirements of time domain applications. To avoid signal distortion other than pure rise time increase (e.g. overshoot, ringing), the group delay has to be largely constant, so the filter types of choice for time domain applications are ones with Bessel and Gaussian type filter responses.

Given the typical high bandwidths needed for present-day digital applications, all filters used here are passive filters (combinations of resistors, capacitors, and inductors).

5.2.4 AC Coupling

In quite a few interfaces, and also as an option in most oscilloscopes, one finds so-called *AC coupling*. What this means is that a capacitor is placed in series with the data line(s), as shown in Figure 44(a). This forms a high-pass filter, which effectively avoids any DC current flow, only AC components above a certain frequency (the filter bandwidth) can get through[59]. There can be several reasons for this. In an oscilloscope this allows to look at a small signal component that rides on a large DC offset (otherwise the large offset

[58] The only common exception to this rule is AC coupling (see section 5.2.4), which constitutes a high-pass filter.

[59] Which is the reason AC coupling is also called "DC blocking" .

would require a rather insensitive range setting which would not resolve the small signal well). Or it may be that the device driver must not be terminated to ground like most benchtop instruments (oscilloscopes, spectrum analyzers) do – CML (current-mode logic) pull-down drivers being a prime example which work well with AC coupling because they provide their own biasing. Or it can be a way to prevent damage to the sensitive I/O circuitry for hot-pluggable daughter cards.

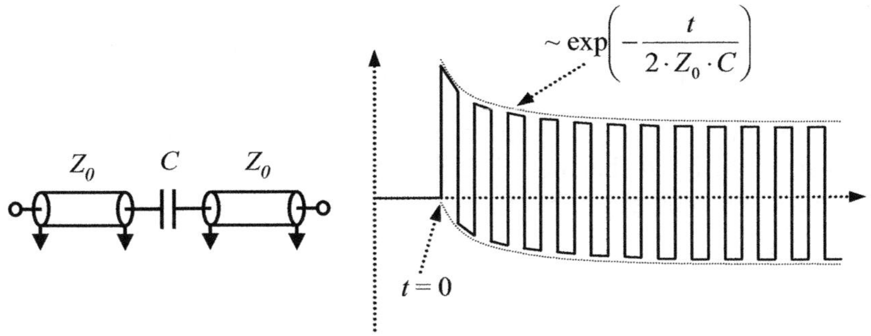

Figure 44: (a) AC coupling of a transmission path. (b) Initial settling behavior ("warm-up") of an AC coupled clock signal.

In order for AC coupling to work, the data stream (and the coupling capacitor) must fulfill certain criteria: First of all the data stream must be DC balanced (i.e. it must spend equal times in the high and the low state) within every not-too-long section of the stream. The figure of merit here is the *maximum running disparity* (RD_{max}) which tells us how many unbalanced "excess bits" (either ones or zeros) can accumulate at most within the data stream. The filter time constant $T_C = 2 \times Z_0 \times C$ must then be large compared to RD_{max} times the bit period to avoid excessive DC wander (change in average level on the output). The maximum wander A_{wander} is given by

$$A_{wander} < A_{signal} \times \frac{RD_{\max} \times T_{bit}}{Z_0 \times C}, \tag{40}$$

where A_{signal} is the signal amplitude, T_{bit} is the bit period, C is the capacitance, and Z_0 the line impedance. In addition, this scheme needs sufficient warm-up cycles to reach a steady state (in the end the output will always be centered around zero if the duty cycle is 50%), as illustrated in Figure 44(b) which shows a clock signal going through an AC coupled path. The settling time constant is identical to the filter time constant T_C. This settling behavior is a frequent cause of "My test run on the production tester fails, but signal looks fine when I check it with a scope" – the test strobes come after an insufficient number of warm-up cycles, but when looking at the scope, the pattern has been looping for a long time. To minimize this warm-up, we should make *C* large enough to avoid droop, but not any larger.

Examples for data streams that will work with AC coupling:

- A clock with 50% duty cycle ($RD_{max} = 1$).
- Manchester-coded data in NR[60] format where each data bit is coded as two bits (either 10 or 01; $RD_{max} = 1$).
- 8b/10b encoded data ($RD_{max} = 3$) which is frequently used in modern serial transmission schemes like SerDes.
- PRBS[61] data of sufficient length (a 2^N-1 PRBS pattern has $RD_{max} = N$, but is not exactly DC balanced because the pattern contains one more "one" than it contains zeros, but for long patterns, around N > 7, this becomes negligible).

Cases where AC coupling will not work are data streams in RZ or RO format[62], patterns where there is no guaranteed DC balance (e.g. data from an A/D converter) or where RD_{max} can be excessively long (e.g. memory address lines).

As an example for how to size the coupling capacitor, assume we have a SerDes data stream at 1 Gb/s, 8b/10b encoding ($RD_{max} = 3$), $Z_0 = 50\ \Omega$, signal swing 400 mV, maximum allowed eye closing 1% = 4 mV:

$$C \geq \frac{A_{signal}}{A_{wander}} \times \frac{RD_{\max} \times T_{bit}}{Z_0} = \frac{400\text{ mV}}{4\text{ mV}} \times \frac{3 \times 1\text{ ns}}{50\ \Omega} = 6\text{ nF}.$$

To be on the safe side (among other issues capacitors can have large variations), let's choose C = 10 nF. We will then need to run warm-up cycles for at least several time constants, let's say 5 (this gives less than 1% residual level error), i.e.

$$T_{warmup} = 5 \times 2 \times Z_0 \times C = 10 \times 50\ \Omega \times 10\text{ ns} = 5000\text{ ns},$$

which corresponds to 5000 cycles (bits)!

[60] NR = non-return, the signal stays at the same level until a bit of opposite polarity is sent.

[61] Pseudo-random bit stream - we will talk more about PRBS in section 9.2.4.

[62] Return-to-zero (one), the signal always returns to low (high) level before the next bit period.

5.2.5 Providing Termination Bias

The inputs of virtually any bench-top test equipment we encounter are either high impedance (slower oscilloscopes for example, or active FET probes) or provide matched 50 Ω termination. In the latter case more likely than not the termination goes straight to ground. This is fine in many cases, but certain device driver architectures (ECL and PECL for example) require that the end of the line be terminated to some other voltage, e.g. the positive or negative supply voltage. They may not even work at all if the termination pulls them towards ground! And they may not be so user-friendly and provide their own bias if AC coupled (like CML drivers would do). So what can we do?[63]

Using a high-impedance passive or active probe may work in some cases (this again depends on the particular architecture of the driver), but this means we leave the receiver end of the line completely unterminated – strong reflections are guaranteed. As long as the driver itself provides matched source termination this is of less concern, but first we cannot always be sure about the quality of its termination (keep in mind that what we want to *test* is the driver, so it isn't good practice to make too many assumptions about its good behavior), second this may not correspond to the actual use in the final application (in case the receiver there does provide termination), so the validity of our test results becomes questionable. Thus something more elaborate is required.

First, we could think of still using the high-impedance probe, but adding some termination circuitry of our own. This is indicated in Figure 45(a). As a side effect, the termination cuts the effective (Thevenin) source impedance in half, so the effect of any probe capacitance is reduced by half as well. (Of course the voltage swing is only half, too). Unfortunately a simple resistor connected to a power supply will not do the trick: The current through the resistor has to change as fast as the signal rises, which can be as fast as a few picoseconds. Clearly no power supply is up to that task, since already the cabling would cause many orders of magnitude more delay than that. So we need to provide some decoupling as well, indicated as a capacitor in Figure 45(a).

Things get even more complicated when the scope input is not high impedance but rather 50 Ω. In this case we cannot simply add some resistor since the termination would no longer be matched. Instead we have to create a "Thevenin equivalent" network that looks like 50 Ω at its input. At the very least this needs two resistors (plus the power decoupling mentioned

[63] We should note that most digital semiconductor testers have adjustable termination voltage on their digital pins, but not necessarily on their analog instruments. So in this chapter we will focus on benchtop equipment, especially oscilloscopes.

before of course); Figure 45(b) shows one possibility. The signal swing at the oscilloscope is attenuated, the attenuation being a tradeoff against the voltage necessary to drive the DC offset, and with the immunity against impedance tolerances of the oscilloscope.

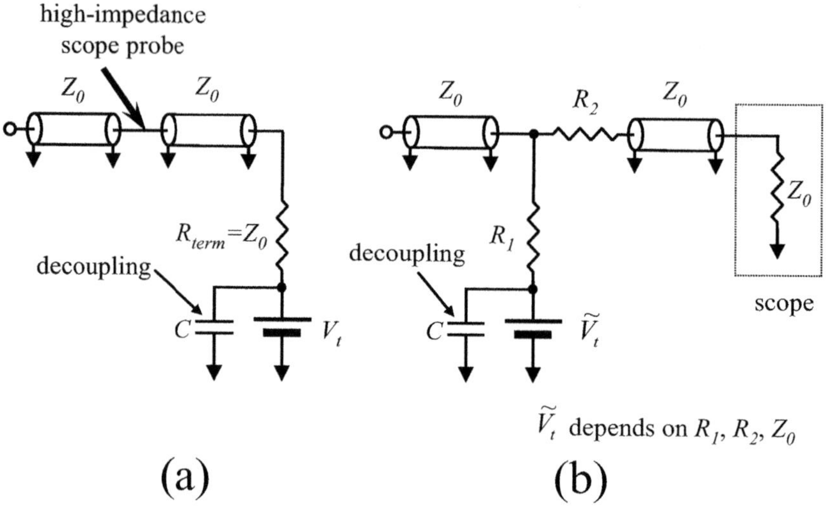

Figure 45: Providing matched termination to a bias level: (a) For a high-impedance probe. (b) Thevenin-equivalent termination for a matched impedance probe that by itself terminated to ground.

Second, we can go out and look for a probe with adjustable termination voltage. As a matter of fact, the selection is not very large, but there are models with SMA inputs where we can either program the termination voltage, or supply it with our own power supply (the probe takes care of the decoupling). This is an excellent and flexible solution with a single downside: Being an active probe, even though one of the best available, its bandwidth is limited to around 12 GHz or less, so we can't use it for faster data rates than maybe 6 Gb/s.

A second very common solution is the use of so-called *bias tees*. These devices are available off-the-shelf from several vendors. Such a bias tee, shown in Figure 46, is a combination of a DC block (a capacitor in series with the transmission path, i.e., AC coupling) and an inductor hanging off the path. The capacitor acts as a high pass filter and blocks any DC current into the oscilloscope. A DC power supply attached to the inductor provides the bias voltage, while the inductor prevents any AC signal to leak out into the power supply. While often used, this solution has several limits and shortcomings:

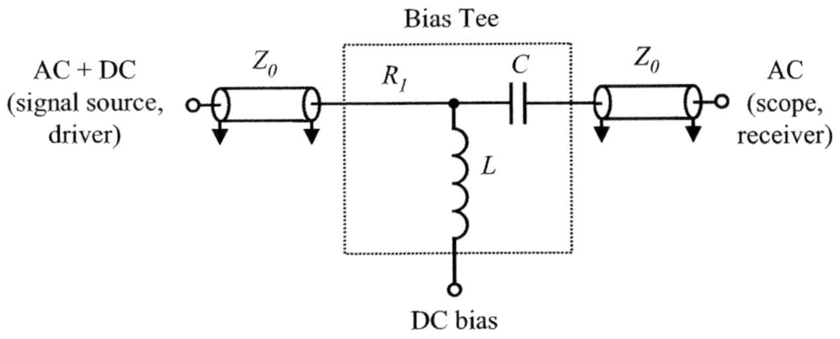

Figure 46: A bias tee is a good way to provide a DC bias to the driver, as long as the signal fulfills certain criteria.

First, the capacitor in the line constitutes AC coupling, i.e. a high-pass filter, so only frequency components *above* a certain limit are transmitted to the scope. Good bias tees have a limit as low as a few kHz. One cannot make the capacitor too large either (to lower the limit) because then its parasitic imperfections will degrade its high-speed performance, and settling will be too long (see section 5.2.4). The inductance shall *block* high-frequency components, so we want to make it as large as possible. In addition the inductance in combination with the capacitance form a resonant circuit with an approximate resonance frequency of

$$f_{res} = \frac{1}{2\pi \times \sqrt{LC}}, \tag{41}$$

which again gives motivation to make L and C as large as possible to lower the operating frequency limit. If we remember that long sequences of ones (or zeroes) correspond to low frequencies, it becomes clear that we cannot transmit really arbitrary patterns; the capacitor will charge up and block, and the inductor will open up and conduct. What's more, real coils are rather bad approximations to an ideal inductor. They have a lot of parasitic capacitance between their turns (which short-circuits the high-frequency components the inductor is supposed to block) as well as non-negligible ohmic resistance that will cause a voltage drop whenever current flows into the device under test – we will need to compensate for that. Practical designs aim to minimize the capacitive parasitics through the use of conical coils (coil diameter increasing from the signal path to the other end) where the first, small, low inductance section blocks the highest frequency components with very little parasitics, and subsequent, increasingly wider sections (thus also with increasing parasitics) block lower frequency components.

Second, since the DC component of the signal is blocked, the waveform on the scope will always be DC balanced around zero. This spells trouble when the waveform's duty cycle (the time it is high vs. the time it is low) is not constant – the whole waveform will move up and down, with a time constant equal to

$$T_c = 2 \times Z_0 \times C. \tag{42}$$

Since such a bias tee contains a DC block, the implications are the same as for AC coupling (section 5.2.4) – namely, the data stream must be DC balanced or at least have constant average duty cycle throughout the pattern (and the duty cycle must not change, lest the output signal levels change as well).

Third, some vendors offer termination blocks with either fixed (PECL, ECL) or adjustable termination voltage. Internally they implement a high-performance version of the power-decoupled Thevenin equivalent network mentioned before. For the driver those terminations look like 50 Ohm to the termination voltage. The limitations are their bandwidth (around 10 GHz is the highest available), and the fact that the output signal is usually attenuated (between 12 and 20 dB, i.e. by a factor between 4 and 10), which means reduced signal-to-noise ratio on the scope.

If our signal is differential (as are more and more high-speed transmission schemes anyway), there is a fourth, even better solution, as shown in Figure 47: First, we assume our lines are uncoupled lines, so even and odd impedance are the same (let's assume 50 Ω cables). We can then implement differential termination in its T-variant (see Figure 19(b)), where the third resistor disappears (because even and odd mode impedance are equal). The midpoint (center tap) voltage never moves, so there is no absolute need for any decoupling (we should still provide one to make the setup more immune against common mode noise). The beauty of this solution is that it retains DC coupling (no warm-up cycles or constant duty cycle necessary) and all parts are available off-the shelf in high-performance versions (20 GHz bandwidth or more), so this is the highest-performance solution available. For the two termination resistors we can (ab-)use an off-the-shelf high-bandwidth power *splitter*. Then all we need in addition are two power *dividers* and a DC power supply. Since the scope inputs draw some current (or, seen differently, the path from the power supply to the scope ground constitutes a voltage divider hooked up to the device driver), we need to set the power supply to some value higher or lower than the required termination voltage.[64] We do need to assure that the true and

[64] The exact voltage to force depends on the input impedance and the termination voltage at the system driver, i.e. on the actual current drawn by the signal source.

complement signal do not have any skew to each other since this would create sudden common mode spikes that we will have a hard time terminating completely[65] even when the center tap capacitor is present, so we need to match all cable delays etc. to a small fraction of the signal rise time.

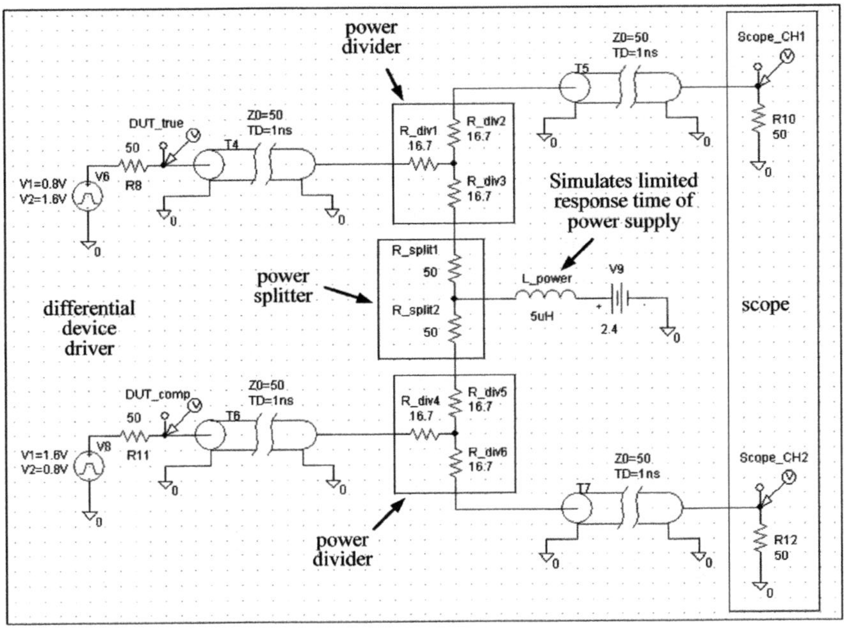

Figure 47: Providing DC biasing for a differential source while retaining maximum system bandwidth, using nothing but off-the-shelf accessories.

5.2.6 Attenuators

Sometimes it is desirable to reduce the signal amplitude to be measured. A frequent case is when using an equivalent-time sampling scope (which in order to maximize bandwidth does not have any amplifiers/attenuators built in before the sampler). If the instrument input is 50 Ω, then a simple resistor in front of the input would do the trick, for the price of an impedance mismatch into the instrument. E.g. to reduce the amplitude into the scope by half, we would place a 50 Ω resistor in front. This constitutes an ohmic 1:2 voltage divider, but now the total input impedance is the sum of the scope impedance and the resistor, i.e.

[65] Even though this termination scheme does provides decent *partial* termination/attenuation of common mode.

$$R_{tot} = R_{scope} + R_{atten} = 50\,\Omega + 50\,\Omega = 100\,\Omega\,,$$

and this would cause strong reflections back into the system under test. This approach would not work at all if the input were high impedance (it would only aggravate the low-pass effect of any parasitic probe capacitance present).

The solution is a three-resistor network as shown in Figure 48. The resistors can always be chosen in such a way that the effective input impedance into each side is 50 Ω *as long as the load on the other side has 50 Ω impedance.* As an example, a -6 dB attenuator (reduces the signal by half) would have $R_1 = R_2 = 16.67\ \Omega$ and $R_3 = 66.67\ \Omega$. This is nothing but the symmetrical power divider we looked at before, but with one transmission line impedance merged directly into R_3. Of course such a network built from discrete resistors would not achieve very high bandwidth because of the significant parasitic capacitances and inductances of real resistors. High-bandwidth commercial attenuators instead employ distributed thin-film resistors and the best achieve bandwidths in excess of 40 GHz.

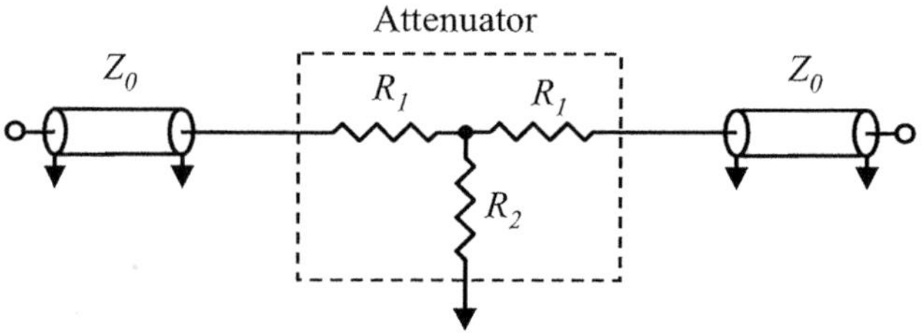

Figure 48: Resistive broadband attenuator.

The design formulas for this attenuator for an attenuation factor of *N* are

$$R_1 = Z_0 \times \frac{N-1}{N+1}, \quad R_1 = Z_0 \times \frac{2N}{N^2-1}, \tag{43}$$

and the attenuation in dB is of course

$$attenuation(\text{dB}) = -\,20 \times \log_{10} N\,. \tag{44}$$

Apart from simple amplitude reduction, attenuators have another important application: They can be used to reduce the impact of mismatched loads: If an attenuator (attenuation factor -x dB) is inserted into the signal path, the transmitted signal is reduced by -x dB, but reflections from e.g. the receiver going back to the source and there getting re-reflected traverse the attenuator two additional times and thus experience -2x dB additional loss – as a result, the signal-to-noise ratio improves by 2x dB. The higher the attenuation value, the better the termination (for the price of reduced transmitted amplitude). In theory on could of course always employ a matching network that would provide *exact* 50 Ω termination[66], but such a network would have to be tailored to each specific load and thus is unlikely to be available off the shelf, while high-bandwidth attenuators are easy to come by.

5.2.7 Delay Lines

Sometimes we need to delay signals by a certain amount of time, but as far as possible without affecting their waveform or their jitter. One application is to look at the signal before a trigger with an analog or an equivalent-time sampling scope. Another frequent situation is the need for a stable, linear, well-defined delay as a timing reference (e.g. we may need to step a strobe signal relative to our measured signal, or to measure and compensate for the time base nonlinearity of an oscilloscope). Finally, a finely variable delay will allow to tightly match the timing of the two signals of a differential pair, which may have picked up some skew e.g. because of slightly mismatched cable lengths.

Only passive delay lines can assure that they don't add random jitter (any active circuit will). In principle a passive delay line is just a fancy name for a piece of low-loss transmission line (low-loss to minimize frequency-dependent attenuation and waveform distortion that would introduce additional data dependent jitter). So if all we need is a fixed delay, we can get that by inserting a piece of cable of appropriate length (use semi-rigid or rigid coax to avoid delay changes that inevitably occur when a cable is flexed and deformed). Achievable delays are in the range of a few 10 ps up to a few 10 ns (corresponding to physical lengths between a few mm and several m); above that length losses are likely to exceed our tolerances.

If we need a series of discrete delay values, we could cut a set of cables with those values and put it the appropriate one at the given time, but this is rather cumbersome (not counting the wear and tear on the connectors) and difficult to automate (if we have to switch the delay size under computer control). A better approach is to use high-bandwidth microwave relays to switch those cable sections in and out of the path. If we size their delays in steps powers of 2 (e.g. 1 ns, 2 ns, 4 ns, 8 ns), we can cover a wide range of delays, equally spaced, with a very limited number of sections. The minimum step size is limited by practical restrictions on the minimum feature lengths and manufacturing tolerances for the delay of the cables,

[66] Funny enough, such impedance *matching* networks are called "*mismatch* pads"!

relays, and solder connections. Such cascaded delay lines are commercially available.

Continuously variable and highly linear delay lines can be built as so-called trombones. They basically consist of two parts of transmission line that slide into each other while keeping good electrical contact. Such trombones can be designed for manual operation (both parts are threaded and so move in or out when turned against each other, or be driven by a stepper motor under computer control. If well designed (good impedance control) such delay lines can be extremely linear because they rely only on accurate geometrical parameters for their delay; building them as airlines minimizes losses so they achieve bandwidth reaching many GHz. And the resolution can be almost arbitrarily small (well below 1 ps). That said, they are still mechanical parts that always have some tolerance and some wiggle, so for maximum repeatability and accuracy one must always approach a given delay setting from the same direction (e.g. always start with a smaller delay and than increase it to the desired value).

Chapter #3

TIMING AND JITTER

6. STATISTICAL BASICS

Statistics plays an important role in measurement. If we want to know how reliable the numbers are that we just measured, find out if a feature we found is really a feature or just a random aberration caused by noise, or want to understand, describe and characterize timing jitter or voltage noise on our signals, statistics gives us the mathematical tools to do so. In the next few sections we will get a short overview on the most important concepts that we need to do meaningful timing and jitter measurements.

To keep the pain to a minimum (most people happen to find dry mathematics painful), we will not attempt anything close to a comprehensive treatment of the statistics of measurements, as this would go far beyond the scope of this book. Instead we will outline a well-chosen handful of concepts that are extremely useful in the everyday work of anybody whose task it is to take measurements and evaluate the results.

6.1 Statistical Parameters

Whenever we take a measurement repeatedly, we will find that there is some variation in the result from one repeat to the next. (If there isn't, then chances are we use some sort of digitalization in our acquisition and the digitizer's resolution is coarser than the actual variation of the result.[67] The

[67] In this case we should really be worried about the systematic (as opposed to random) error introduced by that limitation!

number of measurements can be quite large and difficult to oversee, so statistics makes an effort to describe the whole set with as few numbers as possible. As always in such cases, be aware that describing a large set of data with only few parameters inevitably throws away a lot of information for the sake of simplicity. This works fine in many cases, but it means that even if two sets of data give the same parameter values, they are not necessarily equal (remember that from our rise time discussion?). It often pays to have a closer look at the original data set (e.g. through use of a histogram) to see if the simplification is justified or not.

The parameters that we will use in this book are the mean value $\bar{x}$, the peak-to-peak spread $pk-pk$, and the standard deviation σ.

The arithmetic mean $\bar{x}$ over a set of N measurements x_i is defined as

$$\bar{x} = \frac{1}{N} \times \sum_{i=1}^{N} x_i \tag{45}$$

We see that this mean is dependent on every single item of data, not just on some subset. This is a useful characteristic because it tends to make it very stable for larger sample sizes, since every single contributor has only very small influence. The exception is isolated far outliers that can move the mean by a considerable amount if the sample size is not very large.

In addition to the average value, we may be interested in how wide the variation of the single measurements is. As we can see, the mean of the data sets $\{3,4,5,6,7\}$ and $\{4,4,5,6,6\}$ is the same (namely, 5), but the sets are clearly different. Two figures of merit are commonly used:

First, the peak-to-peak spread is simply the width of the range that the data covers:

$$pk - pk = \max(x_i) - \min(x_i) \tag{46}$$

In the case of our two sets above, this spread would be 4 and 2, respectively. The spread is easy to calculate, but its disadvantage is that it really only depends on two values, the minimum and the maximum. A single outlier can thus completely destroy the usefulness of that number (unless outliers are what we want to catch).

A better parameter is the so-called standard deviation, which gives a measure of how wide the single measurements are scattered around the mean value. It is defined as[68]

[68] Actually there are two possible ways to specify σ. First, weighing it with N, as done in the formula in the text, gives the standard deviation of the actual data set (which is then regarded as the complete set of values). Second, weighing it with (N-1) would give an *estimate* for the "real" standard deviation we would eventually get if we continued to

$$\sigma = \sqrt{\frac{\sum_{i=1}^{N}(x_i - \bar{x})^2}{N}} \tag{47}$$

Because of the square, it weighs larger deviations stronger than smaller ones, so it is not simply the average deviation from the mean (actually, the average deviation of the mean would always be zero by definition and is thus not very useful). The standard deviation is also called the root-mean-square (RMS) deviation.

6.2 Distributions and Histograms

When we perform a measurement, the number of possible outcomes is almost always a finite number, let's call it P. This is either due to a limited resolution of our instrument (e.g. an 8-bit A/D converter can only give 256 possible different readings on its output), or because any differences smaller than some minimum are of no interest to us (e.g. we probably don't care much if the voltage is 1.00000 Volts or 1.00001 Volts), so it is reasonable to count all results that are close together as the same. If we now take a large number N_{tot} of measurements (in statistics, N_{tot} is called the *population*), then we will end up with many groups of the same results. Instead of a long list of all measurements, which would be extremely repetitive and also difficult to comprehend, it is much better to group identical results together and only state the results x_j and the number of times $N(x_j)$ we obtained each particular result, the so-called *frequency* of the result. If we do this graphically, plotting the *bin values* x_j on the horizontal axis, and drawing bars with a height corresponding to their respective frequencies $N(x_j)$ at each bin value, then we get the well-known histogram representation. An example for such a histogram is shown in Figure 49. It could be the result of a repeated measurement of a signal level with an A/D converter. The possible outcomes x_i are often called "bins" because their function of gathering identical (or at least close) results is similar to throwing all results that are the same into the same basket or bin. The distance between subsequent bins is called the width of the bin. In many cases (and all cases we will encounter in this book) all the bins have the same width.

---

acquire an infinite number of samples (the measured set is then regarded as a random sample of that total). However, for increasing N the difference between the two values quickly becomes negligible.

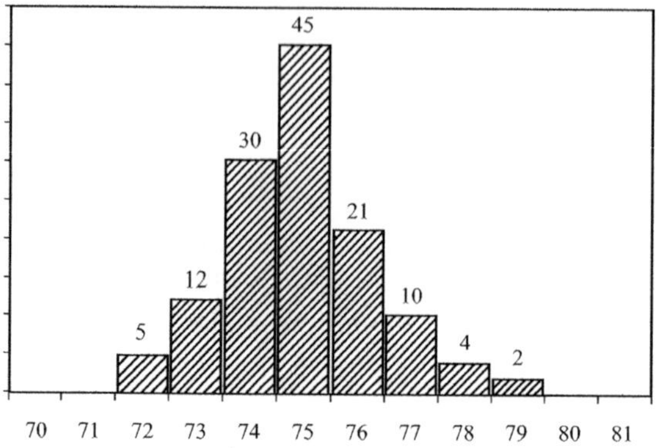

Figure 49: Histogram representation of a discrete statistical distribution.

Using bins and their frequencies makes calculation of mean and standard deviation much easier – remember that otherwise we would have to sum up every single measurement! The total number of measurements is

$$N_{tot} = \sum_{k=1}^{P} N(x_k), \tag{48}$$

and then the mean value is

$$\bar{x} = \frac{1}{N_{tot}} \times \sum_{j=1}^{P} \{x_j \times N(x_j)\}, \tag{49}$$

which looks very similar to the formulas in the preceding chapter, except that instead of summing up all the single results, we multiply each result by its frequency (the number of measurements that yielded this value) and add together all those products. The standard deviation again is the sum of the squared deviations from the mean, and it is calculated as follows:

$$\sigma = \sqrt{\frac{1}{N_{tot}} \times \sum_{j=1}^{P} \{N(x_j) \times (x_j - \bar{x})^2\}} \tag{50}$$

Finally, the peak-to-peak value (also called the *spread*) of the histogram is the value of the largest bin that has a frequency larger than zero, minus the value of the smallest bin that has a frequency larger than zero.

As an illustration, let's do that for the histogram from Figure 49: The total number of measurements is

$$N_{tot} = 5 + 12 + 30 + 45 + 21 + 10 + 4 + 2 = 129.$$

Then the average value is

$$\bar{x} = \frac{1}{129} \times (71 \times 5 + 72 \times 12 + ... + 78 \times 2) \approx 73.94,$$

the standard deviation is

$$\sigma = \sqrt{\frac{1}{129} \times \left\{5 \times (71 - 73.94)^2 + ... + 2 \times (78 - 73.94)^2\right\}} \approx 1.94,$$

and finally the spread of our results is

$$pk - pk = 78 - 71 = 7.$$

6.3 Probability Density and Cumulative Density Function

Probability density function and cumulative density function are two concepts that at first glance can be rather unintuitive when demonstrated on continuous distributions. Fortunately, things get very easy when the distribution is discrete, in other words, when we are looking at a histogram.

The probability density function $PDF(x_j)$ then simply corresponds to this histogram, normalized in height so the sum of all frequencies yields 1 (in other words, the probability that we get *some* result for each particular measurement is of course always 100%). We do this by simply dividing each frequency by the total number of measurements. Those new frequencies then give us the relative probability $PDF(x_j)$ that we get a particular result x_j:

$$PDF(x_i) = \frac{1}{N_{tot}} \times N(x_i). \tag{51}$$

For a continuous distribution, N_{tot} would correspond to the area under the curve. The cumulative density function $CDF(x_j)$ at each point (bin) gives us the probability that we get a result of x_j *or smaller*. We get it by summing up all frequencies of the bins left to the particular bin, plus the current bin:

$$CDF(x_j)=\sum_{k=1}^{j} N(x_k). \tag{52}$$

This corresponds to the total area of the histogram left of x_j (for a continuous distribution, the sum would transform into an integral). To go into the other direction, i.e. calculate the *PDF* from the *CDF*, we simply have to take the differences between subsequent *CDF* values (for a continuous distribution we would have to take the derivative instead of the discrete difference, but all real-world measurement data is discrete anyway):

$$PDF(x_j)=CDF(x_j)-CDF(x_{j-1}). \tag{53}$$

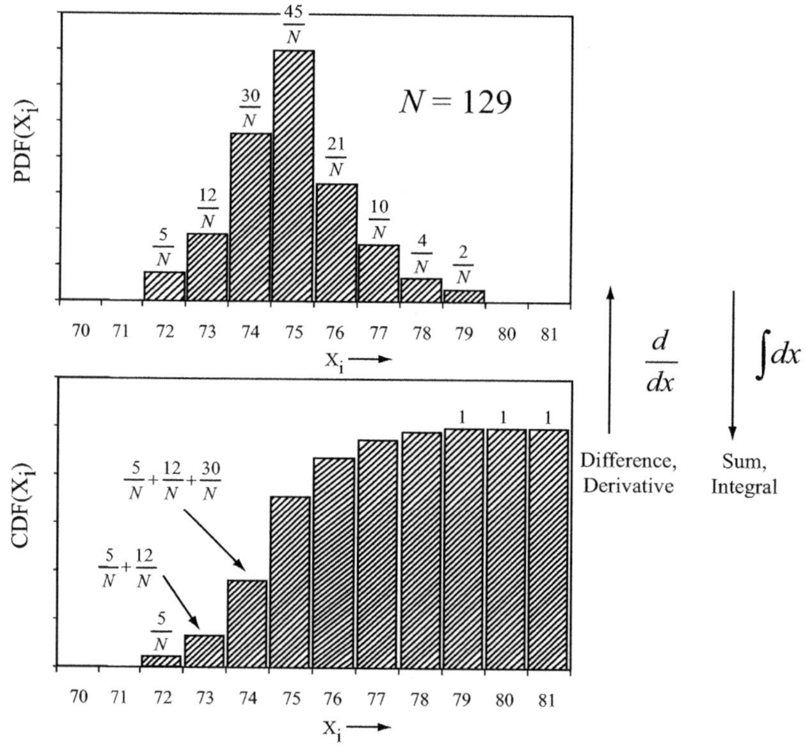

Figure 50: Probability density function (PDF, upper plot) and cumulative density function (CDF, lower plot) for the data from Figure 49.

Figure 50 shows the *PDF* and the *CDF* for the example distribution from Figure 49. As we will see later in this chapter, an important application of the probability density function is the characterization of different types of jitter. Some types of jitter measurements give us only the *CDF* directly, but as we have seen, getting the *PDF* from that is really not a big deal.

6.4 The Gaussian Distribution

6.4.1 Some Fundamental Properties

One of the most important probability distributions is the so-called normal (or Gaussian) distribution. Its general shape is depicted in Figure 51. It is an excellent *approximation* to the *random* variations in a large variety of data sets obtained by measurement. In those cases the peak, coinciding with its mean value $\bar{x}$, is assumed to approximate the real value of the measured property. The variability of the single measurements is given by the standard deviation σ. The beauty of the Gaussian distribution is that it is completely determined by just these two parameters; its mathematical representation is

$$PDF_{Gaussian}(x) = \frac{1}{\sigma\sqrt{2\pi}} \times e^{-\frac{(x-\bar{x})^2}{2\sigma^2}}. \tag{54}$$

The factor comes from the requirement that the area under the curve be unity (i.e. 1), meaning the total probability is 1 (100%), as required for a *PDF*. The formula can be made much simpler by centering on the mean value (i.e. defining $\bar{x}$ as zero on the x axis) and normalizing the scale to σ:

$$PDF_{Gaussian}(z) = \frac{1}{\sqrt{2\pi}} \times e^{-\frac{z^2}{2}}, \text{ where } z = \frac{x-\bar{x}}{\sigma}. \tag{55}$$

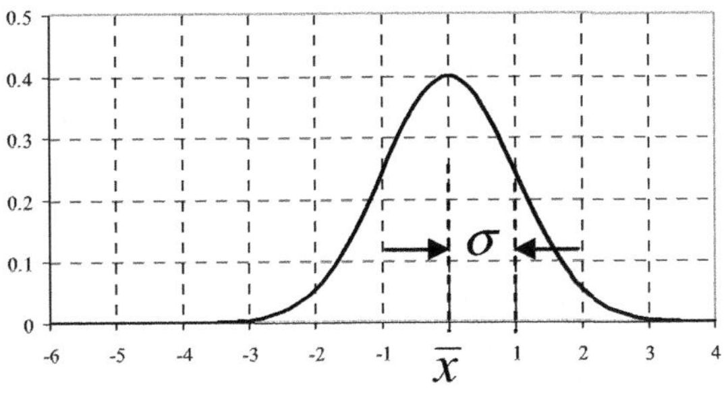

Figure 51: The Gaussian distribution.

After this rather dry introduction, let's get a little more life into this function by looking at how it relates to real-world measurements:

Obviously the bulk of the probability is concentrated near the mean value, but in theory it extends all the way to ±infinity. This means that if we take measurements, while random errors will make the results vary from one measurement to the next, most of the results will be rather close around the mean result. How wide they are spread out is determined by their standard deviation. Large deviations from the mean are possible, but their probability vanishes rapidly with increasing distance. Table 2 gives an idea how fast: It shows the percentage of results that are within a certain number of standard deviations from the mean (this corresponds to the area between $\pm k\sigma$ in Figure 51).

Table 2: Gaussian distribution: cumulative probability for different confidence intervals.

interval	probability
$\pm 1\,\sigma$	0.6827
$\pm 2\,\sigma$	0.9545
$\pm 3\,\sigma$	0.9973
$\pm 4\,\sigma$	$1 - 3.67 \times 10^{-5}$
$\pm 5\,\sigma$	$1 - 4.27 \times 10^{-7}$
$\pm 6\,\sigma$	$1 - 1.973 \times 10^{-9}$
$\pm 7\,\sigma$	$1 - 2.560 \times 10^{-12}$
$\pm 8\,\sigma$	$1 - 1.244 \times 10^{-15}$
$\pm 9\,\sigma$	$1 - 2.257 \times 10^{-19}$
$\pm 10\,\sigma$	$1 - 1.534 \times 10^{-23}$

Let's take for example the interval of $\pm 3\sigma$, which according to Table 2 contains 99.73% of all results. This means only 0.27% or roughly one measurement result out of 400 is expected to lie outside those boundaries. So if we acquire e.g. 10000 samples, we would expect about 27 of them to be outside of those boundaries.[69] Of course there is nothing that would prohibit a result far outside, but as we can see from Table 2, such an outcome is very unlikely.

We can turn this table around and use it as a quick and dirty estimate to see if the distribution we measured is really purely random (Gaussian): Most modern oscilloscopes are able to give those percentage numbers directly, and for large sample size (at least 10000 to make $\pm 3\sigma$ useful) the measured percentages should be close to the numbers in the table. If they aren't, we should start looking for deterministic influences on our signals.

A second important conclusion is that if we take more and more measurements, it will become more and more likely that we get at least some

[69] The probability numbers in above table correspond to the area below the Gaussian *PDF* curve *outside* $\pm n\sigma$ around the mean.

values that are further and further away from the mean, i.e. the spread of our data set is very likely to grow with increasing sample size: While with only 10 measurements it is highly unlikely to have a value exceeding 3σ (we'd expect 0.027 of such measurements, i.e. most of the time there will be none), we saw before that with 10000 measurements we'd expect quite a number (27) of such results. Seen the other way, if we want to catch such outliers, we have to acquire many, many samples. This becomes very important when doing jitter analysis, as any single such event may produce a communication error in the data transmission. The intervals in Table 2 are also called *confidence intervals*: For our purpose it suffices to remember that if we acquire, let's say, at least 10000 samples, we can be reasonably *confident* that we caught all outliers up to 3σ. With those numbers we see that for everyday measurement purposes (with sample sizes between a few and a few thousand measurements), we can safely assume that the observed peak-to-peak spread in numbers will not exceed between six and eight standard deviations.

One further conclusion is that even though they continue to be popular, parameters like minimum, maximum or peak-to-peak spread are poorly suited to describe a Gaussian distribution, or any distribution that contains a random (Gaussian) element. First, minimum and maximum depend on just a single data point each (and hence the spread on just two), even when the distribution itself may be made up from millions of data points. As a result, those three parameters are notoriously unstable and will vary considerable from one repetition of the measurement to the next, in contrast to mean value and standard deviation which depend on all data points together and thus are extremely stable parameters especially for large data sets.

Second, since the Gaussian distribution is theoretically unbounded, we know that if we just continue to acquire long enough, we will get larger and larger values, so in the limit the spread will approach infinity. Thus at the very least a number like the peak-to-peak spread for a given dataset should be accompanied by information about the sample population (i.e. how many data points were acquired) and if available some information about the type of distribution, so the reader can infer how reliable the value is.

What cannot be stressed enough is that the Gaussian distribution, simple as it is, is always only a continuous approximation to discrete sets of actual data, and there are other limitations as well. One is the assumption that its spread is virtually unbounded, even though with small probabilities for large deviations. This is clearly an approximation as in nature there is obviously always some practical upper or lower limit. A length measurement for example will always yield positive numbers, grouped around some mean, since no physical object can have a negative length, which clearly violates the Gaussian distribution's assumption that it extends to minus infinity (albeit with small probability). But in this example, as long as the standard deviation of this data set is sufficiently small compared to its mean value, the error introduced by this idealization is negligible.

Second, the variations between the single measurements may not be purely random, resulting in a distribution that does not look or behave Gaussian at all (more to such distributions in the next section). In layman's terms, Gaussian distributions come about by the cumulative result of many *small, independent, random* contributions. In the length measurement example it may be the small jitter caused by each of our many muscle fibers in our arm holding the yardstick, each independent from the other. But when half of the measurements are taken by us and half by a friend, the overwhelming part of the variations may come from different ways of how he is holding and reading the yardstick compared to us, which is not such a summation of any random small contribution but rather just a single well defined *deterministic* contributor.

6.4.2 How Many Samples Are Enough?

Very often we find ourselves averaging over several repeats of a measurement in the expectation that the mean value will somehow represent the "true" value better than any single measurement, because it will tend to "average-out" all those random errors causes by noise, jitter, limited measurement resolution, etc. But how good will it be?

First, let's be aware that there is no absolute guarantee as such that the mean will in each and every case be close to the true value. As a simple example, the average number of points when throwing a die will be 3.5. But as we all know there are cases, although rare, where one gets ten or more times 6 in a row, so the mean would be 6. Luckily, the larger the number of measurements averaged gets, the less likely this scenario becomes.

Put into statistical terms, the single measurements x_i will have some random distribution with a certain standard deviation σ. We will assume that the distribution is Gaussian, but this is not necessarily always the case. Now let's repeat the measurement N times and calculate the average $\bar{x}^{(1)}$ over those samples, then perform it another N times and calculate a new average $\bar{x}^{(2)}$ and so on. Then statistical theory tells us that those averages $\{\bar{x}^{(1)}, \bar{x}^{(2)}, \ldots\}$ will again have random (Gaussian) distribution, but with much narrower width (i.e. less variation), given by a new standard deviation.

$$\bar{\sigma} = \frac{\sigma}{\sqrt{N}}. \tag{56}$$

So the standard deviation (average deviation from the true value) is reduced by $\sqrt{N}$. In other words, the accuracy of our averaged result increases with the square root of the number of measurements we are averaging over. We can turn this around and, if we know the standard deviation σ of our single measurements and the desired accuracy $\bar{\sigma}$, calculate the number of measurements (the required statistical population) we have to acquire and average over:

$$N_{required} = \left(\frac{\sigma}{\bar{\sigma}}\right)^2 . \qquad (57)$$

Remember that $\bar{\sigma}$ is only a measure for the average deviation of our mean from the true value, but from the section about Gaussian distributions we know that it is highly unlikely that we get a deviation that is larger then a few standard deviations. What we can also see is that the required number of samples increases sharply (quadratically) when the desired accuracy is much smaller than the variation of the measurements. Usually there is a limit to how much averages are practically feasible, either because measurement time becomes prohibitively long, or because the system under test is not sufficiently stable for such long times but tends to drift. It is usually better to improve the quality of our measurement (i.e. reducing σ) instead of trying to hit the problem with averaging over a huge number of samples!

As an example, let's say we measure the voltage of our output on a scope, and the numbers we get vary with a standard deviation of 20 mV (e.g. due to noise on the signal). How many measurements do we need to take to get the mean value with an accuracy of 10 mV (confidence level 99%)? A confidence level of 99% means the peak-to-peak variation must be less than about $\pm 3\sigma$ (ref. to Table 2), so we get

$$N \geq \left(\frac{20mV}{\frac{1}{6}\times 10mV}\right)^2 = 144 .$$

In other words, we need to average over at least 144 values. Would we instead want to be accurate to 1 mV, then the number would go up to 14400!

A word of caution though, this really only works this way if the distribution is random in nature, but can fail miserably for deterministic distributions, especially when the error source is somehow correlated to the measurement process.

7. RISE TIME MEASUREMENTS

Before we jump head first into general timing and jitter – a complex topic – we will have a closer look at another type of timing measurement: signal rise times. It will show quite a number of potential traps that can easily invalidate any rise time result. Again we should remind ourselves that rise time (a time domain parameter) and bandwidth (a frequency domain parameter) are only two sides of the same coin, so limitations or aberrations on the former translate directly into limitations for the latter.

7.1 Uncertainty in Thresholds

Already the seemingly innocent definition of "rise time" is not without problems. Let's stick for example with the 10/90 rise time, per definition the time span that the signal takes to rise (or fall) from 10% of the swing (i.e. 10% above the final level) to 90% of the swing (i.e. to within 10% of the final level). The trouble is that in practice the start level and the final level are never known with absolute accuracy. First, noise on the signal can make the determination of the "real" levels difficult, even though we can always improve that by averaging over more than one repeat of the signal (provided the signal is repetitive). More serious is the concern that we may not even be able to measure those levels at all: The signal may be degraded by losses (skin effect or dielectric loss), or we could have some serious ringing before the signal finally settles to some level. If now the measurement interval is not long enough to exceed this settling interval, we never actually see the level that the signal ultimately reaches. Skin effect is a particularly nasty offender there because its square-root dependency with time means it doesn't go away for a long, long time; on the other hand, we need high timing resolution to resolve a fast rising edge, which puts an upper limit on the length of the interval we can look at. A common mistake connected with this is that one lets the oscilloscope determine high and low level automatically (on most scopes this is even the default because it is seemingly convenient for the user). Don't! When one zooms in on the edge (in the hope of getting a more accurate measurement that way), the scope has no way of "knowing" what the signal will settle to outside the measurement area, and will use erroneous levels without telling us.

How can we get around this? If we are lucky enough to know beforehand what the levels are (e.g. because they are the internal, known voltage rails of a driver we characterize), then it is good practice to dial those numbers as the 0% and 100% levels into the scope measurement setup. Otherwise one can only zoom out far enough to see if and where the signal converges, measure those levels, and again use those numbers hard coded as one's measurement parameters. If we really do want to use the automatic level measurement, then at least turn on the measurement annotations that show what the scope uses as its reference levels. This will let us spot quickly if things are completely off!

7.2 Bandwidth Limitations

We have already seen how the instrument bandwidth BW can affect the rise time seen by the instrument, so this section can be brief. Let us just recall that the bandwidth is inversely connected to the instrument's rise time T_r:

$$T_r = \frac{k}{BW}, \; k \approx 0.33 \ldots 0.60, \tag{58}$$

and that we have to add the squares of all component rise times. Gaussian filters have $k = 0.33$, typical oscilloscopes have $k \approx 0.4$ because their frequency response falls of faster with increasing frequency. For accurate rise time measurements, the system rise time should be at least three times the signal bandwidth, and even this would still be marginal; four to six times is desirable.

An interesting fact is that systems with Gaussian responses will perform worse than systems that have a more "brick-wall"-like drop-off above their 3 dB bandwidth if the signal rise time is slow, but they have advantages when the signal rise time gets into the range of the system rise time. This comes about because a Gaussian response means there is already some considerable attenuation well below the 3 dB bandwidth limit, but on the other hand it still passes some amount of higher frequency components, whereas scope responses (non-Gaussian) are typically very flat all way up to the 3 dB limit but then fall off quickly (see also Figure 27 in section 5.2.3). So a slower rising signal whose fundamental frequency and first few harmonics are all well below the system bandwidth will pass more or less undisturbed; when the rise time gets smaller, thus the frequency components move higher, the partial passing of the Gaussian filter means the signal does not lose as much of its fast-changing higher harmonics.

7.3 Insufficient Sample Rate

Another thing to look out for is the effective sample rate of the instrument used (for real-time scopes this is the time interval between two subsequent sample event; for equivalent-time acquisition, e.g. an equivalent-time sampling scope it is of course the time delay increment from one point to the next.). From the discussion about digital bandwidth we know that the sample rate poses a bandwidth limit by itself, and this in turn means it will increase the signal rise time. This becomes visible when the true signal rise time is of the order of (or smaller) than the time delay between subsequent samples.

As a rule of thumb it follows that without any interpolation (or linear interpolation), the sample rate should be at a minimum ten times higher than the signal rise time to allow for meaningful rise time measurements.

7.4 Interpolation Artifacts

Real-time sampling oscilloscopes use some clever interpolation methods, usually of $\sin(x)/x$ type, to maximize their timing resolution. While very useful in most instances, this is not without problems in the case when the signal rise time gets in the order of the sampling interval. The interpolation then produces preshoot and overshoot on the displayed signal (which are really artifacts not present on the actual signal). The surprising results of this are rise times that can be faster than the real signal rise time! All this is an effect well known in FFT theory called the "Gibbs phenomenon": Theory tells us that it will occur whenever we have significant signal change between subsequent samples. Figure 2 shows graphically the actual signal (no overshoot or preshoot) as well as the reconstructed signal (which has pre-shoot and ringing).

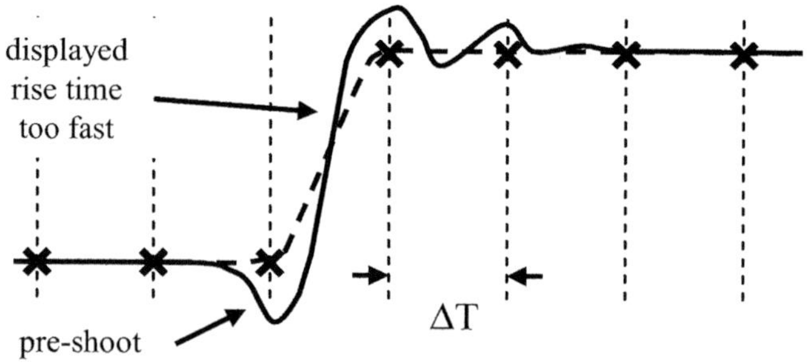

Figure 52: Artifacts caused by sinx/x interpolation (Gibbs phenomenon).

The good news is that interpolation still vastly improves the scope performance – now only a minimum of less than 3 samples is necessary to accurately represent and edge, compared to ten or more without interpolation. Not completely coincidentally this sampling criterion matches the demand based on the Nyquist theorem that the sampling rate be at least twice is high (and in reality, rather 2.5 to 3 times) as the highest frequency in the signal (which is always given by the rise time, not by the edge rate!).

A very useful hint that can keep us from taking hours worth of invalid data: Always look out for this suspicious preshoot on the trace before the transition! Unless we know for sure the signal source we use is producing such a preshoot, or there is a reflection from a previous edge coming back in right before the edge (both are rather rare cases), the chances are very high it indicates insufficient sample rate, and as a result our measured rise times are off. If possible, try to modify the sample rate of the scope and see if the displayed curve remains stable – this is a good indicator that the sample rate is sufficiently high.

If we are already running at maximum sample rate, then the use of an equivalent time sampling scope for comparison can be helpful since such an instrument has almost unlimited equivalent sample rate.

7.5 Smoothing

To make noisy signals look "nice"[70], older scopes used a technique called "smoothing". In a nutshell what they did was calculate a moving average over several sampling points, typically at least 5. Analog scopes did something similar through an analog exponential filter. While this creates traces that look pretty to the eye, it is a really bad idea for high-frequency measurements (and rise time measurements are among them), because it works very similar to an analog low pass RC filter that has to charge up over some time and thus reduces the effective bandwidth of the instrument. Look at Figure 53(a) to see what happens: To determine an *approximate* effective bandwidth or rise time number for this filter, let's assume we send a very fast rising signal into the scope (rise time much smaller than the time interval ΔT between two samples). Now look at Figure 53(b): If the average goes over N successive samples, then it will take the displayed curve N samples to go from low to high, while our input signal was rising more or less instantaneously, so the visible rise time is the rise time of the filter:

$$T_{10/90,\ smoothing} = 0.8 \times N \times \Delta T. \tag{59}$$

The factor 0.8 comes from the conversion between 0/100 rise time and 10/90 rise time. The effective bandwidth of the smoothing process is then given by

$$BW_{smoothing} \approx \frac{0.33}{T_{10/90,\ smoothing}} = \frac{0.33}{0.8 \times N \times \Delta T} \tag{60}$$

[70] in more technical terms: to remove high-frequency noise from the signal to obtain a smoother trace on the screen

Be aware that the formula is only approximate because the filter behavior is not Gaussian at all, but its response is rising linearly; still it is a good rule of thumb. We can also see that the tighter the samples are spaced, the smaller the bandwidth impact of smoothing becomes. For real-time sampling making the sampling interval hits very soon some practical limits because it would require excessive sampling rate, but if one uses equivalent-time sampling, the interval can be made almost arbitrarily small.

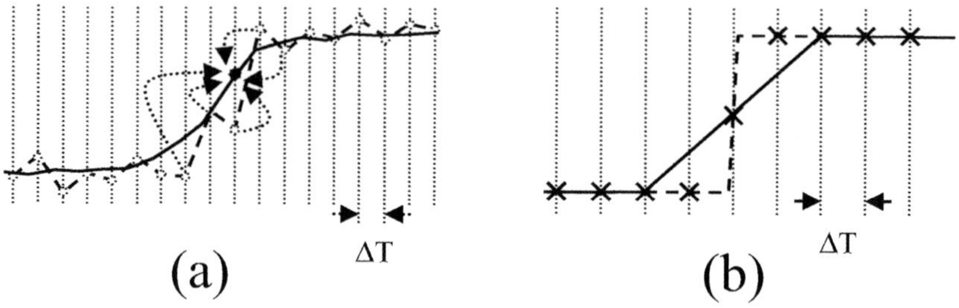

Figure 53: Curve smoothing: (a) Basic algorithm, (b) Original and smoothed signal – the latter shows increase rise time, corresponding to bandwidth limitation by smoothing.

7.6 Averaging

While the bandwidth degradation caused by smoothing is probably not really too surprising, who would think that digital averaging over several traces, available on every modern digital scope, could affect the effective bandwidth as well? In a slightly naïve picture, averaging means taking basically the same curve over and over again, so any random noise will average out, but it should not affect the "real" (average) shape of the curve, should it? In fact, it does. What we forgot to take into account is that virtually every signal has some random jitter (and if the signal itself does not have much, then the scope still has trigger jitter, time-base instabilities, etc.), so the signal trace shows up at a slightly different timing position on each repeat. Let's have a look at Figure 54: For simplicity, it assumes the signal can end up on only two timing positions, T and $T+dT$, it does that with equal probability, and it has an actual rise time (for simplicity we use 0% to 100%) of T_r. Looking now at the average of those two possible traces we see that the rise time of the displayed, averaged trace is increased compared to the single-shot traces.

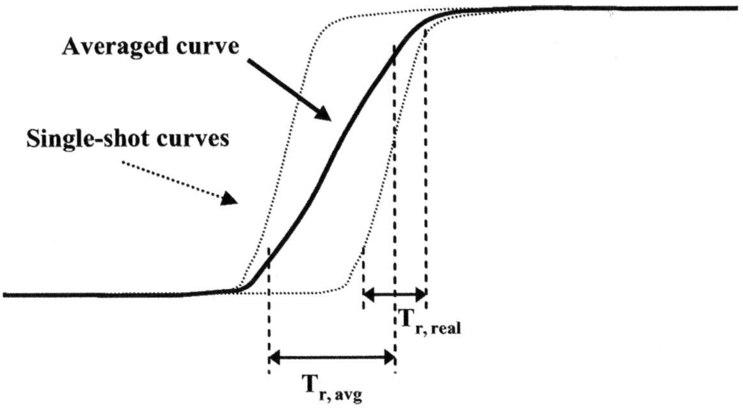

Figure 54: Bandwidth limitation (rise time degradation) caused by averaging of a signal that has jitter.

Again we can express the degradation caused by this effect as a low-pass filter with a certain equivalent bandwidth. In reality most of the time the situation will not be as simple as in our example, where the edge occurs on only two different positions, but there will be a continuous range of possible positions (jitter). Under the assumption that the jitter distribution is Gaussian (we'll talk more about that in the next section), the jitter having a standard deviation of σ_{jitter}, the effective 3 dB bandwidth is

$$BW_{averaging} = \frac{\sqrt{\ln 2}}{2 \times \pi \times \sigma_{jitter}}. \tag{61}$$

As we can see, without jitter ($\sigma_{\mathrm{jitter}} \approx 0$) this bandwidth becomes infinite, corresponding to a vanishing effect on the signal. The equivalent rise time (i.e. the rise time we would see on the screen if we sent in a signal that in reality is rising infinitely fast, but has some jitter) is of course again given as

$$T_{10/90,\,averaging} = \frac{0.33}{BW_{averaging}} = \frac{0.33 \times 2 \times \pi \times \sigma_{jitter}}{\sqrt{\ln 2}}. \tag{62}$$

On real-time scopes there is an easy way around this problem: Instead of taking a measurement on a trace averaged over many repeats, take measurements on each repeat, and average those measurements. The trace on the scope screen will *look* worse because the jitter is not averaged out, but the quality of our results will be better! Sometimes this method may run into trouble though, when the noise is so large that the oscilloscopes picks up the wrong spike as the edge, instead of the desired edge, and thus measures complete nonsense. In this case we could possibly first determine the amount of jitter on the edge (later in this chapter we'll get to how this is done), then turn averaging back on and correct the obtained result:

$$T_{10/90,\,signal} = \sqrt{T^2_{10/90,\,measured} - T^2_{10/90,\,averaging}}\,. \qquad (63)$$

On an equivalent-time sampling scope a trace is always a mixture of a large number of acquisitions (normally just one sample per acquisition is taken), so there we will always get hit by this phenomenon. But fortunately the effect is rather small for realistic cases: Let's assume a data stream of 10 Gb/s (i.e. 100 ps bit interval), with a rise time of $T_{10/90}$ = 50ps and jitter of 2 ps RMS. The rise time is chosen just small enough for the signal to approach full level within the bit interval, and the peak-to-peak random jitter must of course be small compared to the bit interval. For a bit error rate of no more than 10^{-12} this peak-to-peak jitter is 14 times the RMS value, i.e. 28 ps our case, which is already a very noisy, marginal design (real-world 10 Gb/s data streams are more likely to be well below 1 ps RMS). The averaging-limited bandwidth is then

$$BW_{averaging} = \frac{\sqrt{\ln 2}}{2 \times \pi \times 2\ \text{ps}} \approx 66\ \text{GHz}$$

and the equivalent rise time is

$$T_{10/90,\,averaging} = \frac{0.33 \times 2 \times \pi \times 2\ \text{ps}}{\sqrt{\ln 2}} \approx 5.0\ \text{ps},$$

which is very small compared to the signal rise time of 50 ps and so won't hurt our measurement.

8. UNDERSTANDING JITTER

8.1 What Is Jitter?

Jitter is defined as the short-term deviation of a signal's transition time from its ideal position in time.[71] This means that an event (for example a

[71] Longer-term changes (i.e. timing changes that happen slowly, at a low frequency) are usually referred to as "wander" instead of "jitter". A commonly used limit is 10 Hz, but

transition from low level to high level) is supposed to take place at some time t but actually happens at time $(t+\delta)$. What that "ideal position" is depends on the system under consideration as well as the way jitter is defined in a specific situation. It may be the timing relative to some system clock, or it may be a certain time distance from some previous event. What's more, there are different very common situations we are interested in when looking at the timing of a system, for example:

- Jitter generation, i.e. what is the jitter of the signal produced by some source when compared to a highly stable (ideal) reference source. This is an important number when looking at a clock source or a pulse generator.
- Jitter insertion, i.e. how much jitter does a certain component in our signal path – e.g. an amplifier – add. In other words, what is the jitter of the component's output signal relative to its input signal.
- Jitter transfer, which is very similar to the previous item, but is usually connected to clock recovery circuits or phase-locked loops. Here we want to know – given a jittery input signal – how much of that jitter makes it to the output. Since a phase-locked loop can follow timing variations as long as they don't happen too fast, it will block low-frequency jitter and transfer only higher frequency changes. In contrast to jitter insertion here the total jitter at the output is normally smaller than the jitter at the input.

8.2 Effects of Jitter – Why Measuring Jitter Is Important

8.2.1 Definition of the Ideal Position

Since we said jitter is the deviation from some "ideal" position, a very valid question is what this ideal position is. This really depends on the architecture of the particular system we are looking at.

8.2.1.1 Data Stream with Separate Clock

The easiest case is when the data stream comes with a separate clock (strobe), provided by the data source together with the data. By definition this clock signal is our ideal reference – everything else in the data stream is

---

this is by no means a natural law but will depend on the specific system under consideration.

supposed to happen relative to this clock. It clearly needs at least two lines (and normally much more) – one clock line and one (or more) data lines.

The ideal position is given by the edges of this clock signal, usually either directly by the edges (if the transmission scheme uses edge-centered strobing), or by the middle of the interval between two adjacent clock edges. It also depends on the particular transmission scheme if only one edge polarity (either rising or falling) or both polarities are used to clock in the data.

To take measurements we can then go and use this clock signal as our trigger of the oscilloscope or as the "Start" edge source for a time stamper. On a real-time sampling scope another option is to use two channels and acquire the clock signal on one of them and the data signal on the other, and do the timing calculations as post-processing on the data. Triggering in this case could be completely asynchronous to the data stream.

Semiconductor production testers on the other hand are often not prepared to handle this type of clocking[72]. They usually assume that all timings are driven by themselves and the device under test will follow the global timing. There is only a handful of digital channel cards available (usually on high-end testers) that are geared towards this. In all other cases the best we can do is again to acquire shmoos[73] of clock as well as data and then post-process them to get the relative timing, which is clearly limited to repetitive effects and stable edges, we can't really acquire any instantaneous clock-to-data relationship.

8.2.1.2 Clock-Less Data Stream

A second, not very common case is that we don't have any actual clock signal at all. Instead the implicit assumption is that things happen at regular intervals, whose length we may or may not know. The venerable legacy RS-232 protocol, used on every PC's serial port, is one example – and it will only work if both the sender and the receiver are set to the same data rate. In this case the first transition defines the start of the data stream; it triggers the acquisition start in the receiver, which then strobes at regular intervals given by the data rate. One way to see this is that the receiver (e.g. a scope or a tester if we live in the device testing world) has its own, stable reference and compares incoming events to that time base, i.e. the receiver's timing system defines the ideal timing.

On the other hand if we measure such a data stream on an oscilloscope, there is not necessarily a need to know the data rate beforehand. Instead we

[72] In the world of large-scale testers we speak of "source-synchronous strobing".
[73] More to shmoos later (in section 10.2).

can use mathematical fit algorithms to determine the average data rate as well as each edge's deviation from the ideal timing thus determined.

8.2.1.3 Embedded Clock – Clock Recovery

Recent transmission schemes do away with any separate clock altogether, and instead "encode" the clock directly into the data stream. This is done by guaranteeing that the data has always a sufficient number of transitions, i.e. there are no long continuous sequences of zeros (or ones). Usually the data has to be encoded to achieve this. This way the receiver has a realistic chance of figuring out the data rate of the data stream, no matter what it is exactly (within certain limits of course, and in practice those limits – the so-called *capture range* – may be very narrow). The receiver can then lock a phase-locked loop (PLL) onto the data stream that gets re-synchronized whenever a new transition arrives. The PLL thus re-generates the clock that is needed to latch the data into the receiver. This whole process is called "clock recovery". A schematic picture of such a receiver is shown in Figure 55.

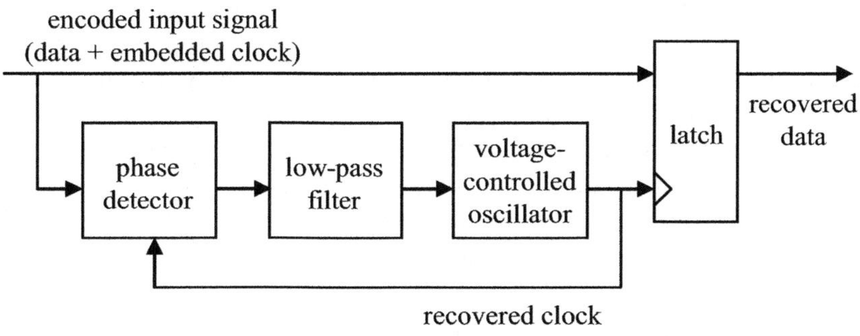

Figure 55: Digital receiver with clock recovery (for serial data with embedded clock).

The big advantage of this scheme is that if the data stream has some not-too-fast[74] timing variations or timing drift, then the receiver will follow them as long as those variations are slow compared to the loop response time of the PLL[75]. The ideal position in this case is given by this recovered clock[76].

[74] i.e. spanning many bit periods or transitions

[75] Typical limits for the maximum frequency response of a fast PLL are a few MHz for data rates in the Gb/s range.

[76] We could consider the previous case of a clock-less data stream a very restricted case of embedded clock, where the frequency capture range is almost zero, i.e. no clock timing drift exceeding a fraction of the bit interval is allowed.

At the same time this is a major headache for anybody who wants to characterize the jitter with an external instrument, e.g. an oscilloscope or a production tester. This is because those instruments usually want to compare everything against their internal time base (the "implicit clock" of before), so they are ill fitted to "listen" to the clock embedded in the data stream. There are several solutions to this problem:

Sometimes the system under test offers an output of the recovered clock, so it can be used directly to trigger the scope. If such an output is not available then one can build an external clock recovery circuit that gets fed with a portion of the data signal. It will recover the clock just like in the "real" system, which then can be supplied as a trigger to the oscilloscope. Unfortunately there is usually no guarantee that this circuit behaves exactly like the circuit in the real receiver. Its loop bandwidth, stability, settling time, jitter, and many more details may be different. If the loop bandwidth is smaller than the real system, then it will not follow faster changes that the real receiver would still catch, and if it is faster, it will compensate for fast changes that will cause the real receiver to fail. And it is very inflexible – the loop response may be difficult (or impossible) to change. On the other hand it is a very fast solution what regards acquisition time since it avoids any processing overhead, and it works for many different types of instruments including equivalent-time sampling scopes, as well as production testers as long as they offer source synchronous strobing. Many recent high-end oscilloscopes offer this hardware clock recovery as a built-in or external option with loop parameters corresponding to standardized specifications of certain transmission standards (so-called *golden PLLs*).

Secondly one could capture the data stream with sufficient oversampling (i.e. several samples per bit period), and then do clock recovery by post-processing the signal in software. Of course for this to work the instrument must be able to capture to full data stream in a single shot, a requirement that disqualifies all but real-time sampling scopes, and it puts high demands on the achievable sample rate and the available capture memory size. On the upside this solution is as flexible as one wants it to be. It is very easy to implement any desired PLL response characteristic in software, including any golden PLL, and one can even change the characteristics after the measurement has been taken. Also, in contrast to hardware clock recovery options, it does not add additional jitter (a "software-PLL" does not have jitter). The price for this flexibility is increased processing time, which becomes more of an issue when the data streams become very long.

8.2.1.4 Edge-to-Edge Jitter vs. Edge-to-Reference Jitter

So far all timing values (and thus the timing jitter) on a signal were referenced to a separate reference signal (either supplied externally or generated from the signal). A different approach is to reference the edge timing to some preceding edge, *N* bits before the edge under consideration. This way of jitter referencing is often called "edge-to-edge", as opposed to "edge-to-reference". Actually in most cases this not done because there is such a thing as a "golden edge"; rather it is an easy way to measure because only a single signal is needed. Time interval analyzers often perform this type of analysis because they lack the ability to capture long streams of edges together with a separate clock signal. Real-time scopes can do either one – if we trigger them on the clock, it yields "edge-to-reference", if triggered on the data signal itself it yields "edge-to-edge". Edge-to-edge measurements are particularly popular on clock signals since here we are guaranteed to have an edge each cycle. On the other hand for a data stream, if looking at edges *N* bit periods apart, the edge-to-edge method will completely miss all the edges that were not preceded by an edge exactly *N* cycles earlier. This is clearly not a very desirable situation in most applications.

The commonly reported value in edge-to-edge measurements is the period jitter or the *N*-period jitter distribution, respectively. This is the distribution of the timing interval between an edge and the following or the *N*-th edge after, respectively. As noted this measurement makes most sense for clock signals, much less so for data signals. Another measure is the cycle-to-cycle jitter, which is the difference between the one period and the next. This is a useful parameter if the receiver uses a PLL for clock recovery: In this case the exact value of the period length is of less importance (because the receiver will adjust its clock to match the incoming clock), but sudden variations are of concern since the PLL needs some time to adjust lest it will be thrown off (i.e. loses synchronization).

In the edge-to-reference situation, the jitter measure is the so called "time interval error", shown in Figure 56. The advantage is that the time interval error is just as easy to define for clock signals as for data signals since we have an external reference and do not rely on the presence of a transition in the previous bit interval.

Further problems occur frequently when one has to correlate measurements done in this "edge-to-edge" fashion to other measurements done "edge-to-reference". For example, the edge-to-edge method is largely blind to long-term variations of the edge positions, since most of the time it is applied to adjacent edges or at least edges close together. On the other hand, it tends to overestimate short-term effects: If one edge is too early by an amount Δ_1, the next too late by Δ_2 (always referenced to the clock), then the true maximum jitter is max(Δ_1, Δ_2), while the edge-to-edge method would yield $\Delta_1+\Delta_1$. Obviously the edge-to-edge method can overestimate this jitter by up to a factor up of two.

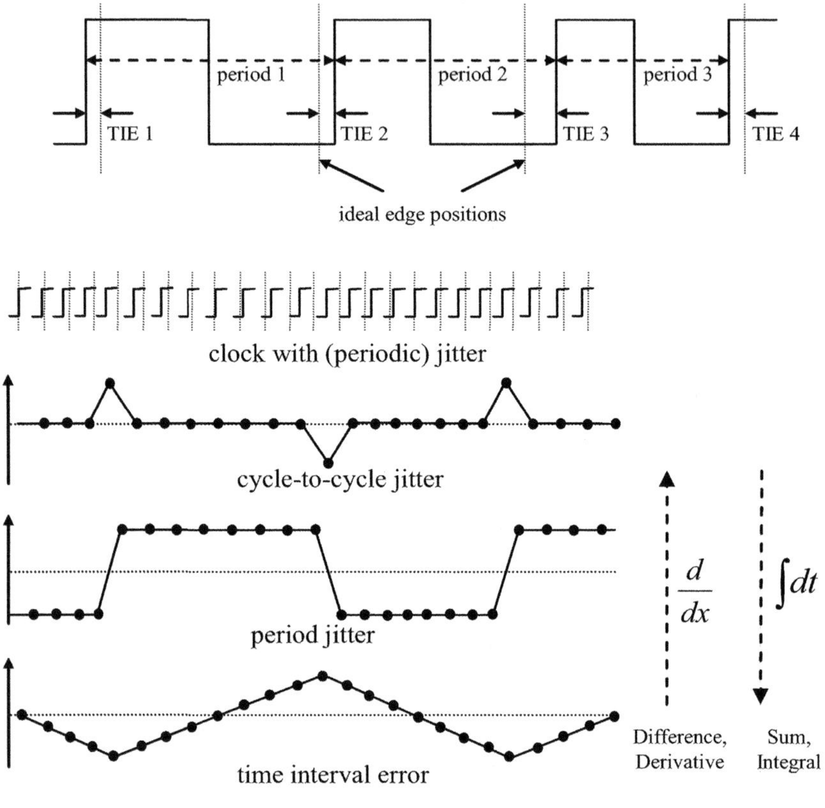

Figure 56: Different ways to define and display jitter, and the resulting jitter trend plots.

8.2.1.5 Jitter Trend

Let's say we have acquired the full data signal (or at least each edge position) in a single shot, e.g. with a real-time sampling scope. We know now how to calculate jitter parameters like period jitter, cycle-to-cycle jitter, or time interval errors – all this will give a series of jitter numbers, one value per edge. This is called the *jitter trend* and is easy do display graphically (as jitter vs. time, or jitter vs. cycle).

Figure 56 shows how the jitter trends of those three types of measurements (period jitter, cycle-to-cycle jitter, and time interval error) are related to each other: The cycle-to-cycle jitter is basically the difference series of the period jitter, while the time interval error is the sum of the period jitter.[77]

8.3 Jitter Types and Jitter Sources

Before we continue to look at other ways to visualize and characterize the jitter on a signal we need to look at the different jitter components that can be present. This will make it much clearer how to read those jitter representations.

In the following sections we will have a closer look at all the different sources where timing jitter on a signal can come from, and what the characteristics of each influence are. Figure 57 gives a breakdown of the different jitter types that we may encounter.

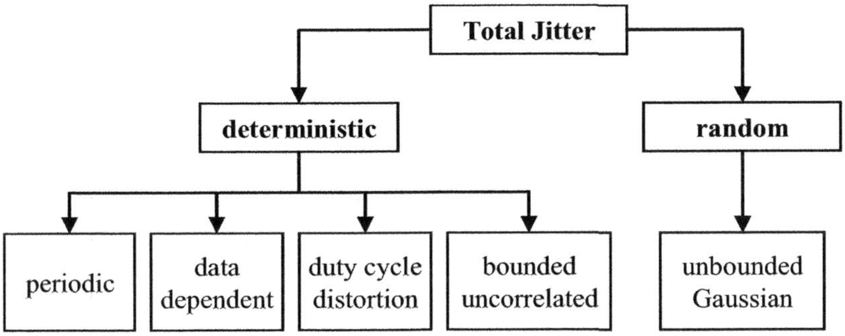

Figure 57: Jitter family tree.

8.3.1 CDF and PDF

If we work with jitter measurements, we need a good way to display and characterize the jitter we find on a signal, either on a particular transition (edge) or on a group of transitions. What we want to convey in the description of jitter is "how likely is it that the edge occurs, let's say, more than 100 ps after its ideal position", because this is directly related to the likelihood that our data transmission will produce an error.

Mathematical theory deals with jitter as a statistical phenomenon, and as such characterizes it using the so-called "Cumulative Density Function" (CDF) and the "Probability Density Function" (PDF). While those terms may sound daunting to the uninitiated, they are actually quite easy to grasp.

[77] If we are looking at a large number of edges, so that the series of period jitter numbers more and more resembles a continuous function rather than a series, those relations approach the derivative and the integral of the period jitter, respectively.

Assume we have a signal source that produces a data stream consisting of a single transition, where the edge has a certain amount of jitter (random or other), and that pattern is repeated over and over again. A receiver samples (strobes) this data once in every repeat. For simplicity let's also assume that the jitter will make the signal transition early exactly as often as it will make it transition late (or, seen inversely, let's define the ideal position as the median position of the transition). If we place the sampling strobe exactly at the ideal position, then the jitter will cause the edge to occur *before* the strobe in half of the cases – in those cases we will receive the correct bit. In the other half of the cases the edge will occur too late – i.e. after the strobe – and we will receive a wrong bit. So the CDF at this position is 0.5 – meaning that in *half* of the cases the transition occurred *at or before the strobe.*

If we move the strobe earlier, then it becomes more and more unlikely that the edge has a chance to occur before the strobe – the CDF decreases. For a very early strobe the CDF will finally go to zero. On the other hand, moving the strobe later and later, the signal will more and more likely have transitioned before the strobe, i.e. the CDF finally reaches 1 – if we just wait long enough, the signal is bound to transition eventually, even with a lot of jitter present. Figure 58 plots a possible shape of the CDF for a given signal. An important property of the CDF is that it is always *monotonous* – the probability that a signal has already transitioned can only increase with time.

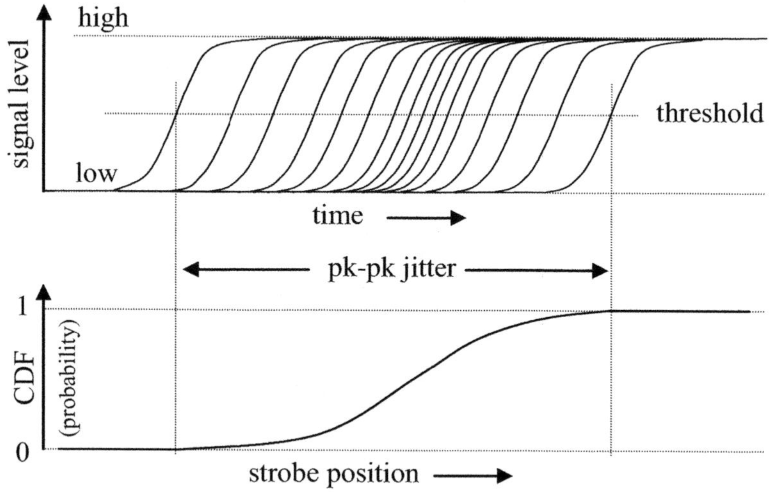

Figure 58: A signal with timing jitter (top) and its corresponding cumulative density function (CDF, bottom).

One very important thing to mention (because it often leads to confusion in everyday work) is that even though a CDF plot looks very much like a signal edge (transition) itself, *it is not a signal edge.* The CDF simply gives the *probability* that the signal has already crossed the threshold at a given instant. On the other hand, there is no "probability" in an edge plot – it shows what the signal *amplitude* is at any given instant, without any probability attached to it. This also means that the rise time of an edge has no relation to the time span it takes the CDF to go from 0 to 1: e.g. for a fast edge the rise time is very small, but if it has a lot of jitter then the CDF rises very slowly, or, in the inverse case even a slowly rising edge can have a fast rising CDF (meaning it has very little jitter).

For many people that are more practically oriented, the CDF looks like a rather non-intuitive monster. What they will ask is "how likely is it that the signal transitions at a specific point in time", e.g. 100 ps after the ideal position. Well, mathematically that probability is exactly *zero* – because a single timing instant means an infinitely small span of time, so the transition is very unlikely (more exactly, infinitely unlikely) to hit this point. But what we can do is give a probability that the signal transitions between, let's say, $t_1 = 100$ ps and $t_2 = 101$ ps after the ideal position, i.e. in a 1 ps range.

Now if the signal has transitioned between t_1 and t_2, obviously this means it has transitioned before t_2, but it had not yet done so before t_1. The likelihood for the first is $CDF(t_2)$, while the probability that it has already transitioned before t_1 (and thus can't transition anymore between t_1 and t_2) is $CDF(t_1)$. Putting both together, a transition between t_1 and t_2 has a probability of

$$p(t_1, t_2) = CDF(t_2) - CDF(t_1) \tag{64}$$

If we now make that range smaller and smaller, we get:

$$\lim_{t1 \to t2} p(t_1, t_2) = \lim_{t1 \to t2} \left(CDF(t_2) - CDF(t_1) \right) = \left(t_2 - t_1 \right) \times \frac{d}{dt} CDF(t), \tag{65}$$

i.e. we eventually get into the situation that the transition probability in this range becomes proportional to the width of the range – e.g. the signal is twice as likely to toggle in our 1 ps range (between 100 and 101 ps) than it is to transition in a 0.5 ps range at the same position (e.g. between 100.25 and 100.75 ps). It makes now sense to express this as a probability density function, the derivative of the cumulative density function:

$$PDF(t) = \frac{d}{dt} CDF(t) \tag{66}$$

In everyday terms the PDF simply means "how likely is it that the signal transitions at time t, compared to some other instant", as long as we keep in mind that this makes only sense (the real probability is only larger than zero) if we look at a short but finite time interval. A high value of the PDF at some instant t means the CDF at this point has a steep slope – it becomes rapidly more likely that the signal has transitioned if we proceed further in time. The width of the PDF gives us the range where we can expect to see the transition happen, and it is identical to the span it takes the CDF to go from 0 to 1.

8.3.2 Random Jitter (RJ)

Already this first item on our list is a frequent source of misunderstanding. Basically "random" means something like "different each time", and so far this is correct – it is a variation from one repetition to the next of the measurement of the same edge. But in addition the usual assumption is that it follows a Gaussian distribution, i.e. meaning for example that it is unbounded – if one collects enough measurements, one will observe any arbitrary large deviation, albeit with fast decreasing probability.[78]. As we already know, a Gaussian distribution is completely determined by two numbers – the average $\bar{x}$ and the standard deviation σ. The former is the "true" edge position, while the latter is a direct measure for the amount of random jitter, in other words, we can characterize random jitter with just a single number[79]. The PDF of random jitter is then given by the standard Gaussian bell curve formula (defining $\Delta t = 0$ as the center of the distribution):

$$PDF_{random}(\Delta t) = \frac{1}{\sigma\sqrt{2\pi}} \times \exp\left(-\frac{\Delta t^2}{2\sigma}\right). \tag{67}$$

[78] There are of course random sources of jitter that don't do that. If there are considerable nonlinearities in the transmission path they can cause the random jitter to deviate from Gaussian behavior (this is a common situation in optical transmission schemes, less so in electrical systems). Crosstalk from an independent source is another example – while it may look completely random over time, it is nevertheless limited in size to some maximum it will never exceed – this type of jitter is usually treated separately and named "bounded uncorrelated jitter" or "uncorrelated deterministic jitter", and we'll talk about it in section 8.3.9.

[79] Sometimes one still encounters random jitter specified by a peak-to-peak value. From the earlier discussion of the Gaussian distribution it should be clear since this distribution is in principle unbounded, this only makes sense when at the same time the confidence level (such as 10^{-12}) is specified.

Random jitter is usually caused by the common influence of a large number of very small, independent contributors – we'll see examples for electrical signals below. The fact that it is (at least theoretically) unbounded is a very unpleasant property since we will never be able to observe the largest possible deviation within a finite measurement period. This already shows how important it will be to characterize jitter so we can make extrapolations based on rather short observation times.

Jitter that does not behave Gaussian but instead is bounded in size is called "deterministic". "Bounded" means there is some largest possible deviation and we will never encounter more than that.

There are a series of possible sources for random jitter, but at their root they are all caused by some sort of noise on the signal.[80]

8.3.3 Noise Creates Jitter

So far we have always dealt with either noise (i.e. deviations in the voltage domain) or jitter (i.e. the deviations in the time domain), but not both at the same time, so we could easily be misled to believe that these two are independent realms. This is not true, as the following section will make clear. While it is at least theoretically possible to obtain a signal that has some jitter but does not have any noise (other than during the transition intervals, that is), the opposite is not attainable.

How does this come about? Let's have a look at Figure 59, which shows a signal transition without any noise present, and the same signal with some noise added.[81] As long as we are concerned with timing, the only thing of interest to us – and to any digital receiver that gets this signal as its input – is when the signal first crosses a given threshold. The noise-free signal will do that at a specific instant, and when there are no other jitter contributors, this timing will be the same every time the measurement is repeated. On the other hand, if there is noise present, then the signal level around the trigger instant will be modified (either increased or decreased), and the signal will cross the threshold either earlier or later than without the additional noise present. In the displayed example, the noisy signal crosses the threshold too early.

[80] For a digital device, unless the noise is noise of the last stage (the output stage) of the device, we will not see it directly, but rather its indirect effect – jitter. This is because each stage in a digital device refreshes the levels – one can consider each stage as a high-gain amplifier that immediately rails. An analog (linear) device on the other hand, for example an op-amp with limited closed-loop gain so it stays in its linear region, will indeed show the (amplified) noise on its output, which again translates into jitter.

[81] In this section, "noise" denotes any aberration from the ideal voltage level, so it comprises not just random noise, but also level changes due to ringing, reflections, or external influences.

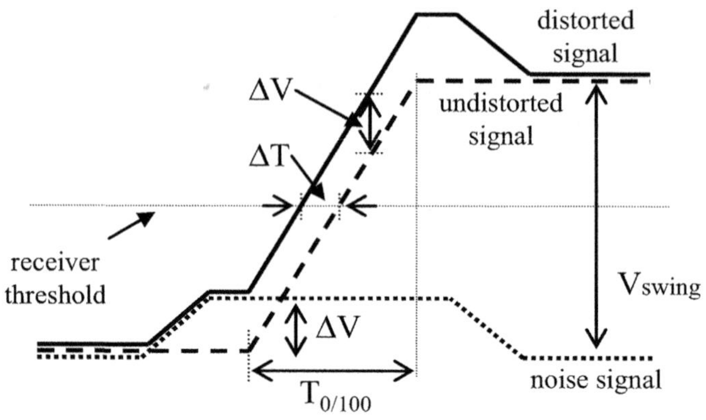

Figure 59: Level noise always causes timing jitter.

Knowing this, the next step is now to determine how much jitter is added by a given amount of noise. We can get a good estimate from Figure 59, where we assume that the edge is roughly linear around the threshold (this is the case for the cast majority of "well-behaved" edges that we encounter in reality). Simple geometric considerations based on this drawing lead to the formula

$$\Delta T \approx T_{0/100} \times \frac{\Delta V}{V_{swing}} \tag{68}$$

In other words, the jitter caused by the noise is simply the rise time divided by the signal-to-noise ratio. Note that as an exception the rise time here is the 0% to 100% rise time, and we assume that the edge is a simple linear ramp. To use the standard rise times $T_{10/90}$ or $T_{20/80}$ we can substitute

$$T_{0/100} \approx 1.25 \times T_{10/90}\text{, or } T_{0/100} \approx 1.67 \times T_{20/80}\,. \tag{69}$$

As an example, say we have a signal with 1 V amplitude and a rise time $T_{10/90}$ of 100 ps. If there is 100 mV of noise present (which may simply be some ringing left over from the previous transition), then the jitter caused by this is of the order of 12.5ps. A more in-depth analysis would of course use the slew rate around the trigger level, but this number is very often hard to obtain while rise times for a device output or an interface can be typically found in the datasheet. Since the slew rate is normally steepest in the middle of the edge, using the 10/90 or 20/80 rise time will give a conservative (slightly pessimistic) estimate for the effect of the noise.

An interesting conclusion from above equations is that the timing of signals with shorter rise times is less affected by noise than for a slower rising signal. On the other hand, steeper rise times are normally used to drive faster data rates, with correspondingly smaller margin for timing errors, so the relative effect tends to stay about constant. Also, faster rise times cause larger reflections at parasitics in this path – roughly inversely proportional to the rise time (see section 1.6.3), so in this specific case rise time reduction would not result in any gain regarding jitter.

From above considerations – the amount of timing shift being proportional to the voltage shift – it is clear that the nature of the noise will translate directly into the nature of jitter induced by it. If we have random noise, we will get increased random jitter, and the result of the timing measurement will change every time it is repeated. If on the other hand the noise is deterministic in nature, e.g. due to reflections, the distribution of the added jitter will be bounded.[82]

Another important conclusion is that if we can reduce the amount of noise either on our signal or in our measurement setup, we will automatically reduce the jitter as well. From the measurement point of view this points us very strongly to the importance of having a very clean, noise free environment in order not to add jitter to our measurement results. This can be achieved in a variety of ways – using low-noise scope amplifiers, reducing reflections through good impedance matching and proper termination, assuring maximum usable amplitude of the signal to be measured so the receiver (sampler) noise is less dominant, etc. This is especially important for the signal that we use as a trigger (in those cases where a separate trigger is needed), because for many measurement situations the trigger jitter will enter – and degrade – the timing measurement result directly. Clearly having a high signal-to-noise ratio on the trigger signal helps, as does using a square wave shaped signal (i.e. fast rise times) instead of a sine-wave shaped signal whenever the former is available, because, as we have seen, steeper edges are less susceptible to noise-induced jitter than more gradually rising ones.

In fact, the most common source for random (Gaussian) timing jitter is indeed noise, even though it may sometimes not be obvious – the signal one is looking at may seem to have very little noise yet it may have substantial random jitter. This can occur e.g. when some component in the beginning or middle of a signal chain has noise (creating jitter). Subsequent *digital* stages will conserve the timing jitter, but produce clean, noise-free levels between

[82] In the case of reflections or waveform aberrations (overshoot, ringing), this will mean we will get the same amount of timing error whenever the preceding stream of data before the particular transition is the same – i.e. those effects produce data (pattern) dependent jitter.

transitions on their outputs. In other words, they will only react to noise close to the transition (input threshold crossing).

8.3.4 Noise Types and Noise Sources

So now we know that even when we only deal with *timing* measurements, noise is an important factor for us. In principle we can denote as "noise" everything that affects the signal level on the line. A rough categorization divides noise into two groups:

First, there is extrinsic noise, caused by something outside our signal path. This can be crosstalk from adjacent lines, feed-through from switching power supplies, ground loops, cosmic radiation, 50 Hz (or 60 Hz) line noise, etc. There is really no general rule for them, as their presence/absence and their size completely depend on the specific system we are looking at.

Second there is intrinsic noise, i.e. noise generated by something in our signal source, signal path, or measurement device. We will now take a closer look at those sources, because as we will see this type of noise follows well-defined rules. Noise in our test setup and instruments puts a lower limit on the achievable measurement accuracy, and better knowledge will enable us to optimize our test setup – remember, voltage noise always translates into timing jitter, so reducing the noise in our test setup will reduce the measured jitter. Furthermore, the specific statistics of a particular noise source transfer directly into the statistics of the timing jitter it causes.

There are four basic types of random intrinsic noise: Thermal noise (Johnson noise), shot noise, 1/f noise, and burst noise (popcorn noise). Let's deal with them one after the other. An important remark beforehand, since all those noise sources are random, we must not add them linearly, but use the RMS value. E.g. 5 mV of thermal noise and 10 mV of shot noise add up to

$$\sqrt{5^2 + 10^2} \approx 11.2\,\text{mV}$$

of total noise.

8.3.4.1 Thermal Noise

Thermal (or Johnson) noise is produced by the random thermal motion of charges carriers (electrons, and for semiconductors also holes), and it affects active as well as passive devices – in other words, even a simple resistor in our path contributes thermal noise. The RMS noise voltage across a resistor is

$$\sigma_{V,thermal} = \sqrt{4kTR \times BW}\,, \tag{70}$$

where k is Boltzmann's constant (1.38×10^{-23} Joule/Kelvin), T is the absolute temperature (in Kelvin; room temperature is approximately 293 K), R is the resistance, and BW it the signal (or instrument) bandwidth. It has a Gaussian distribution, and it is a type of "White Noise", i.e. it has the same noise power per unit bandwidth at any frequency – in other words, its noise power spectrum is a flat line[83] (power is proportional to the *square* of the voltage).

What can we learn from this formula? First, thermal noise gets worse with increased temperature, not too surprising given the nature of this type of noise. So we can reduce it by lowering the temperature of the setup, but this is usually not a real option unless we want to put our oscilloscope into a cryogenic cooler. More important for us are the dependency with the resistance and with the bandwidth.

A lower resistance means lower thermal noise signal. So – apart from possible signal integrity issues if we do not terminate our line – a 50 Ohm scope input yields a 140-fold improvement over a high-impedance 1 MΩ input (this could also be a high-impedance active probe).

Second, the total noise increases with bandwidth, so an oscilloscope with 50 GHz bandwidth inevitably sees 7 times as much of this type of noise as a 1 GHz bandwidth instruments. This can put a serious lower limit on the noise performance of our scope. What we can learn from this bandwidth dependency is that we should not use an instrument with excessive bandwidth (meaning more than maybe 6 to 10 times the signal bandwidth) because this means excessive measurement noise. For example if we measure a signal that will later go into a 1 GHz bandwidth receiver, using a 50 GHz oscilloscope will cause us to greatly overestimate the received jitter because the scope picks up much more ($\sqrt{50} \approx 7$ times more) thermal noise than the receiver will.

Finally, let's have a look at some typical numbers. First, let's say we have a 100 kΩ active probe with a bandwidth of 5 GHz (we neglect for the moment that as we know due to the probe's parasitic capacitance its impedance will be less than 1 kΩ at the highest frequency). The thermal noise in this probe will then be

[83] To be exact we should mention that the formula above would indicate that if we make the bandwidth arbitrarily high, the total noise power would increase to infinity – this is known in quantum physics as the "Ultraviolet Catastrophe". Fortunately for us this is not the case in reality, instead the power rolls off at high frequencies so the total noise power is finite. But at room temperature the formula is valid to well beyond 1000 GHz, i.e. this is of no concern to us.

$$\sigma_{V,\,thermal} = \sqrt{4\times 1.38\times 10^{-23}\times 293\times 10^{5}\times 5\times 10^{9}}\ \mathrm{V} \approx 2.8\,\mathrm{mV\ RMS}.$$

Keep in mind that the peak-to-peak number is several times the RMS value, let's say ±7 times for standard communication links (for a bit error rate of $\approx 10^{-12}$ – see Table 3 in section 9.1.1). If we look at a rather slow signal with 200 ps rise time and 200 mV amplitude, the corresponding RMS jitter caused by this noise is

$$\sigma_{T,\,thermal} \approx T_R \times \frac{\sigma_{V,thermal}}{\Delta V} = 200\,\mathrm{ps} \times \frac{2.8\,\mathrm{mV}}{200\,\mathrm{mV}} = 2.8\,\mathrm{ps\ RMS}.$$

The peak-to-peak jitter is then approximately $2.8\times 14 = 39.2\,\mathrm{ps}$. If we can use a 50 Ω input instead, the noise will be greatly reduced and thus be almost immeasurably small. On the other hand, if our 50 Ω sampling head has 50 GHz bandwidth this would partially negate the improvement by the reduced impedance.

As a final rule of thumb, the thermal noise created at room temperature in a 50 Ω resistor is $0.9\,\mathrm{nV}\times\sqrt{BW}$.

8.3.4.2 Shot Noise

Shot noise is caused by fact that electrical current consists of moving electrons (or holes), thus the electrical charge is quantized (it cannot change in arbitrarily fine increments). Shot noise is created by the random emission of electrons (e.g. from a photo-cathode) or by the random passage of charge carriers (electrons or holes) across a potential barrier, e.g. a semiconductor gate. The effect is especially noticeable when the total current is very small, i.e. only few electrons flow. Then it makes a measurable difference if by chance at some instant e.g. 1000 instead of 1001 electrons arrive per second. The quantized nature of shot noise makes that it its probability statistics is *not* Gaussian, but obeys the so-called Poisson statistics, although at higher currents this statistics quickly approaches the Gaussian one. As a consequence this is one of the rare cases we are likely to encounter where we can observe non-Gaussian random jitter.

The magnitude of shot noise is given by

$$\sigma_I = \sqrt{2\times e\times I\times BW}, \tag{71}$$

where e is the electron charge (1.60×10^{-19} As), I is the average total current, and BW is again the bandwidth. Note that the *relative* noise σ_I / I decreases with the square root of the current – out of the same statistical reason as the relative error of a mean value decreases with the square root of the measurements averaged!

We can substitute voltage V and resistance R for the current and get

$$R = \frac{V}{I} = \frac{\sigma_V}{\sigma_I} \Rightarrow \sigma_V = \sqrt{2 \times e \times V \times R \times BW}\,. \tag{72}$$

This dependency on the bandwidth and the resistance is the same as for thermal noise, so the same considerations apply. Just like thermal noise, shot noise is white noise, i.e. has a constant power density (at least up to a certain – but high – limit).

Let's have again a look at some typical numbers, taking the example from before (200 mV signal amplitude, rise time 200 ps, 100 kΩ probe with a bandwidth of 5 GHz). The instrument's shot noise then yields

$$\sigma_{V,shot} = \sqrt{2 \times 1.60 \times 10^{-19} \times 0.2 \times 10^{5} \times 5 \times 10^{9}}\ \text{V} \approx 5.7\ \text{mV RMS},$$

or roughly double the thermal noise for this particular case. The resulting peak-to-peak jitter of around 80 ps is substantial.

Note that this is only the *instrument* jitter, and it gets added to the "real" signal jitter. The signal itself can have much larger jitter than this if the signal source's internal signal path contains sections with lower current where shot noise becomes more prominent, and that then gets amplified in later driver stages of the path.

8.3.4.3 1/f Noise

The two previous noise sources both resulted in white noise, i.e. constant noise *power* per frequency interval (power being proportional to the square of the voltage, the noise *voltage* increases only with the square root of the frequency). In contrast to that, 1/f noise power density decreases with frequency, so it mostly affects the low end of the frequency spectrum – or, in the time domain, causes slower-speed drift and jitter, although smaller and smaller design scales in semiconductor devices continuously increase the 1/f noise frequency limit.

1/f noise occurs very widespread in things as different as semiconductors, earthquakes, radioactive decay, traffic flow and so on, and there isn't a generally accepted theory that would explain all those cases. In semiconductors the cause is believed to be charge trapping at dislocations and faults. There isn't too much a test engineer can do to reduce this type of noise, other than to choose instruments that are designed well to minimize it. That said, also resistors exhibit 1/f noise: Wirewound and thin-film (metal) resistors have lowest 1/f noise, thick-film resistors have moderate 1/f noise, and carbon film resistors as well as carbon composition resistors are the worst and should be avoided in maximum-accuracy measurement setups.

8.3.4.4 Burst Noise (Popcorn Noise)

Burst noise manifests itself as sudden jumps in the signal to a new level. It is thought to be caused by lattice defects or by microscale beta shifts, but the processes are not completely understood. Most of the noise power is concentrated at rather low frequencies (less than 1 kHz), but since the jumps are step-like, some Fourier components go up to higher frequency ranges.

8.3.5 Periodic Jitter (PJ)

Periodic jitter denotes timing errors that repeat in time (or along the bit stream), but are usually not correlated to the data rate or clock frequency. The frequency of repetition is always lower than half the data rate.[84] Due to its periodic nature it lends itself nicely to treatment in the frequency domain, which is why most of the times a sinusoidal jitter trend is assumed. In case the real jitter is not sinusoidal, it can always be decomposed into a discrete Fourier series and each component can then be treated separately.

Sinusoidal jitter in the time domain produces a distribution given by the following formula (defining time zero as the center of the distribution):

$$PDF_{periodic,\,sinusoidal}(\Delta t) = \begin{cases} \dfrac{1}{\pi\sqrt{a^2 - \Delta t^2}} & |\Delta t| \le a \\ 0 & |\Delta t| > a \end{cases}, \tag{73}$$

where $2a$ is the peak-to-peak width of the periodic jitter[85]. This distribution – with and without added random jitter – is displayed in Figure 60(a) and (b), respectively. It resembles a bimodal distribution but in reality is continuous over the full range. Purely sinusoidal jitter would have a distribution that asymptotically goes to infinity at its edges, but random jitter and our limited

[84] This does not mean the cause of periodic jitter can only have frequencies lower than half the data rate. But since the transitions can in some sort be seen as "sampling" the jitter – with a sample rate equal to the bit rate – higher frequencies alias into the low frequency band limited by half the sample rate, just like we have seen for the Fourier transformation. The aliasing can be seen as a mirroring at the Nyquist frequency. E.g. periodic jitter at exactly the bit rate will affect each edge the same, in other words, it aliases down to DC.

[85] Readers familiar with theoretical mechanics may recognize this formula, as it is the probability density function of the classical harmonic oscillator.

measurement resolution always blur the distribution and produce peaks of finite height and finite width.

If more than one periodic component is present then the distributions of those partial components will of course convolve together. The picture will be different depending if the components are coherent (which normally means they are caused by a single source) or not (in the case we have several independent sources), and the total timing distribution can become quite complicated. This is one of the cases where analysis in the frequency domain can greatly simplify our work.

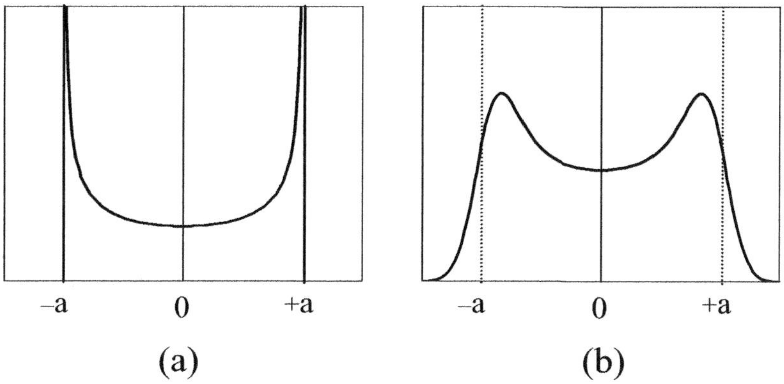

Figure 60: Jitter distributions (PDF) for sinusoidal periodic jitter: (a) pure sinusoidal jitter, (b) with added random jitter present.

Periodic jitter can have a variety of causes – basically anything that exerts a periodic influence. First, it can be the data source itself. A typical example is a N-to-1 serializer (a device that takes a parallel data stream N bits wide and translates it into a serial – one bit wide – data stream). Such devices frequently have some amount of subharmonics at the bit rate divided by N – which simply means particular bits of the parallel input will be offset in time by a constant amount. High-speed digital data sources (data generators or drivers in large-scale automated test equipment) are also prone to such effects because often internally they have to multiplex together several slower data sources to obtain the required high data rate, and each source may behave slightly differently. All devices that use a PLL (phase-locked loop) to generate a fast internal clock locked to a slower externally supplied one may exhibit periodic jitter at the slow clock frequency.

Finally the influence may be external, and here the group of possible culprits is large: Bleed-through from the power supply is a common cause, as is ground bounce, crosstalk from some clock line, or electromagnetic interference from an external radio station or cell phone.

8.3.6 Duty Cycle Distortion (DCD)

The name "duty cycle distortion" is very closely related to clock signals (i.e. regular streams of high-low-high-low...). "Duty cycle" itself is the average fraction of the time the signal spends in the "high" or "low" state, respectively (those two numbers are distinguished as "positive duty cycle" and "negative duty cycle"). Of course it is also possible to assign a duty cycle number to a data stream when one averages over long enough sections of the stream. A duty cycle of 50% means the signal spends equal times in the high state and the low state.

In a simple non-return-to-zero (NRZ) data stream "duty cycle distortion" means that the (average) position of the rising edges is different from the (average) position of the falling edges. In a clock signal this results directly in a non-50% duty cycle, hence the name. Duty cycle distortion is often also called "pulse width distortion" since it makes the duration of a positive pulse different from the duration of a negative pulse of the same number of bits.

Duty cycle distortion can have several sources. The most common are threshold level offsets and differences in the rising and falling edge characteristics.

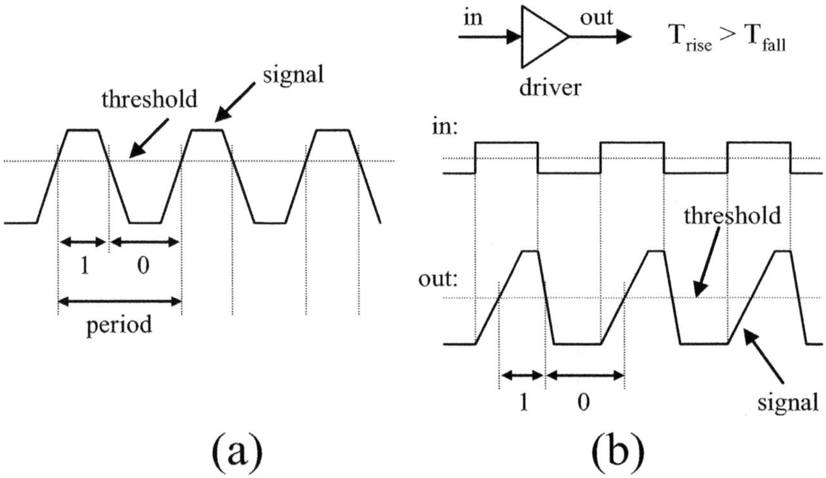

Figure 61: Duty cycle distortion (a) caused by an offset threshold, (b) caused by mismatch between rise time and fall time.

As shown in Figure 61(a), if the threshold level is not the center of the waveform, duty cycle distortion is an almost automatic result. So if one encounters duty cycle distortion, the first thing should be to check if the level threshold is set correctly. Of course it is also possible that it is the signal itself that is offset, or that the ground is shifted in a system employing single-ended signaling[86], which all may indicate a real problem with the signal source. The size of the duty cycle distortion caused by this effect depends on two factors, namely the level offset and the slew rate (or rise times). For a given level change the resulting timing change is indirectly proportional to the slew rate. In other words, signals with faster rise times are less susceptible to threshold level uncertainties.

A second possible source is a difference in the driver behavior for rising and falling edges (most importantly, differences in rise time). Figure 61(b) illustrates such a case: While the driver stage receives all its drive strobes at the correct times, the fall time is faster than the rise time (a common scenario e.g. for pull-down drivers with passive pull-up), and as a result the rising edges lag behind.

Duty cycle distortion yields a bimodal distribution consisting of two sharp peaks of equal height[87], as shown in Figure 62(a), unless one separates rising and falling transitions in the measurement. Theoretically those peaks are Dirac delta functions[88] with infinitely small width and infinite height (but finite total area), but in practice random jitter and our limited measurement resolution always produce peaks of finite height and finite width. In the presence of random jitter (which is unavoidable) this can thus result in a picture very similar to periodic or data dependent jitter if the random component is larger than the duty cycle distortion, so separation of measurements on rising and on falling edges is well worth the effort if at all possible in the specific setup.

The analytic formula for duty cycle distortion is the sum of two delta functions:

$$PDF_{DCD}(\Delta t) = \frac{\delta(\Delta t - a)}{2} + \frac{\delta(\Delta t + a)}{2}, \tag{74}$$

where $2a$ is the peak-to-peak width of the duty cycle distortion. Like for periodic jitter, our limited measurement resolution as well as any random

[86] Differential signaling is largely immune to ground shifts because those shifts cancel out in the receiver which only looks at the difference between the signals.

[87] Equal height because the number of rising and falling edges in a data stream can never differ by more than one – a digital signal cannot for example rise and then rise once again without having fallen in-between – so for any sufficiently long data stream they are virtually identical.

[88] A thorough discussion of the Dirac delta function would exceed the scope of this book. For our purposes it is sufficient to think of this function as an isolated peak, with infinitely small width, infinite height, but nevertheless a finite area of unity (i.e., 1). The integral of the Dirac delta function is the Heavyside step function that is zero left to the transition point and one right to it.

jitter on an actual signal will blur those peaks so they end up with finite height and finite width (see Figure 62(b)).

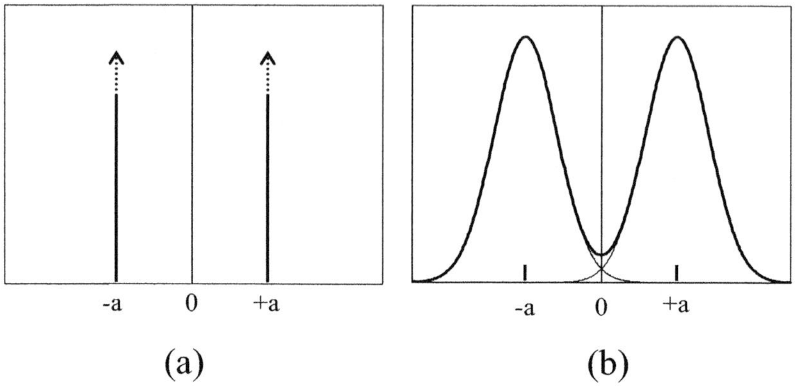

Figure 62: Jitter distribution (PDF) for duty cycle distortion: (a) ideal (pure duty cycle distortion), (b) peaks broadened due to added random jitter.

8.3.7 Data Dependent Jitter (DDJ, ISI)

Data dependent jitter describes timing errors that depend on the preceding sequence of data bits – in other words, the "history" of the data stream. Two sources are the most common causes of this type of jitter:

First, each real signal takes some time to reach (settle to) its final level. Thus, if a transition follows very closely the preceding one, the signal has not yet reached full swing and begins the transition with a "head start". As a result it reaches the threshold level too early, in contrast to a transition that comes after a long transition-less time (where the signal has long settled out), which will then come later. This is illustrated in Figure 63. The rise time is of course dependent on both the driver's signal rise time and the rise time of the transmission path. Since rise time and bandwidth are just two sides of the same coin, this illustrates a very important principle: Any bandwidth limitation automatically results in data dependent jitter! This is the basic reason why one strives to maximize the bandwidth of the transmission channel in order to minimize timing errors.

Second, if we have reflections in the path (due to impedance discontinuities or parasitics) and those reflections happen to arrive just at the time the signal is close to crossing the threshold level, those reflections add up with the "real" signal and may help (make it cross earlier) or hinder (make it cross later) the signal depending on their polarity, and thus affect the timing (which is nothing else than the point in time the signal crosses the threshold).

In a digital data stream (and assuming the transition medium is linear, which is virtually always the case) each transition creates the same set of reflections, and all those partial reflections add up. The total effect on a subsequent edge thus is dependent on the preceding pattern[89]. So again optimization of the transmission path – in this case good impedance control and the minimization of all parasitics – is a very powerful means of reducing timing errors.

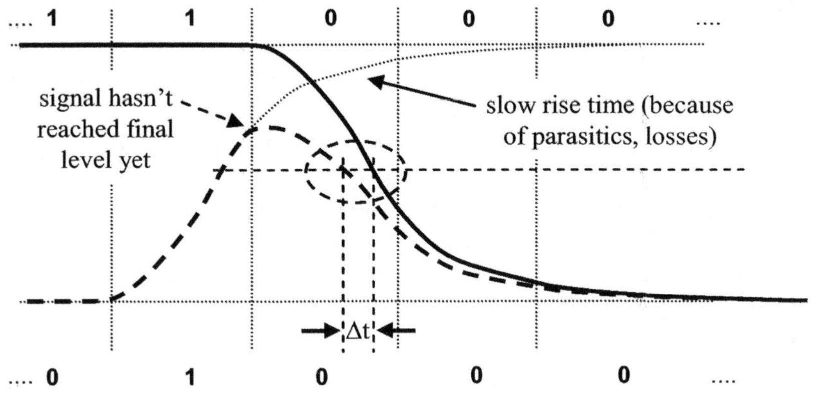

Figure 63: Data dependent jitter caused by rise time limitations.

Overall the mechanism is practically identical to what we discussed in the "noise creates jitter" section (8.3.3) before – any aberration in signal level automatically results in an aberration in timing, the conversion factor being dependent on the rise time (or slew rate).

Since there is always only a limited number of different possible patterns in a data stream of limited length, data dependent timing errors always produce a discrete timing jitter spectrum, theoretically consisting of two or more delta functions:

$$PDF_{pattern\ dependent}(\Delta t) = \sum_{j=1}^{N} \{p_j \times \delta(\Delta t - t_j)\}, \quad \text{where} \sum_{j=1}^{N} p_j = 1. \qquad (75)$$

In this formula, N is the number of distinct sub-patterns, p_j is the probability of the particular pattern occurring, and t_j is the timing displacement of the edge following this pattern.

89 In fact it is of no importance if the signal distortion is a result of reflections or if it is something that comes from the signal source itself – the result is indistinguishable. Of course the former will change if the transition path changes, while the latter will not.

Of course just as discussed for duty cycle distortion, limited measurement resolution as well as random jitter (and other jitter sources) will blur this picture. In this case if the discrete distribution is very dense (denser than the width of the random jitter), the total distribution will no longer be discrete since the peaks will flow into each other. On the other hand only a short preceding pattern section tends to be of any importance (because settling and reflections rarely persist longer than a few bit intervals), which limits the number N of discrete cases that are of practical importance. Figure 64 illustrates a typical distribution caused by purely data dependent jitter, as well as the same distribution, but with added random jitter (which of course is the practically important case). Unlike duty cycle distortion the heights of the peaks are not necessarily equal since there is no guarantee that all sub-patterns occur with the same frequency in the data stream.

Data dependent jitter is also known under several different labels, like inter-symbol interference (ISI, a term commonly used when emphasis it put on the data content of the signal), pulse pulling (because preceding transitions or pulses tend to "pull-in" the timing of subsequent ones when settling time is the cause), data dependent timing errors, or pattern dependent jitter (common in the testing field), which can confuse the novice, but it is really always just the same thing.[90]

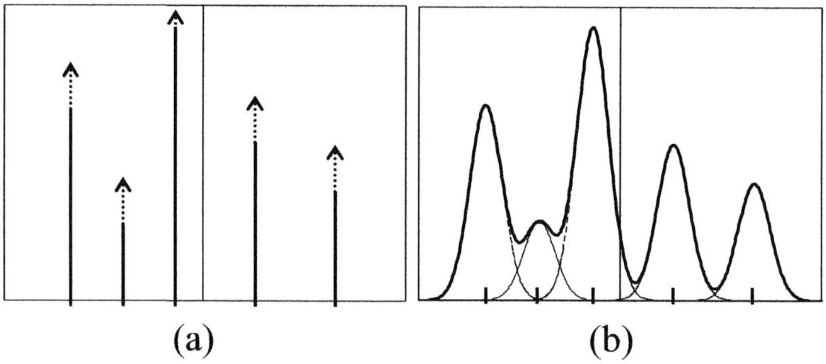

Figure 64: Typical jitter distribution for data dependent jitter: (a) ideal distribution, (b) peaks broadened by additional random jitter.

[90] That said, there seems to be a trend to use the term "ISI" to denote the effect of the data history on the signal as a whole, including e.g. overshoot and other aberrations, while "DDJ" denotes the effect on the transition timing, but many people still use the terms interchangeably.

8.3.8 Duty Cycle and Thermal Effects

Despite their similar names, duty cycle effects and duty cycle distortion are two very different types of jitter.

Depending on the design of the signal source (driver), the power consumption of the driver circuit may be different depending on whether the driver is in the high or the low state, respectively. This is especially true for bipolar designs. FET-based architectures on the other hand are more dependent on the operation frequency, i.e. how many times they transition in a given time interval, since the only time they draw significant current is when switching. The temperature of such circuits will thus also depend on the time the driver has been powered up and the pattern has been running. Different power consumption results in different die temperature, which in turn may affect the bias point of the circuit. The end result is a dependency of the edge positions on the duty cycle of the data stream. While one could in principle roll this into the data dependent jitter effects, it is usually kept separate because the root cause is quite different, as is the typical time constant: While other data dependent effects disappear after no more than a few round trip times at the worst (which amounts to maybe a few nanoseconds), the time constant for thermal effects is much longer, on the order of microseconds.

The timing distribution due to duty cycle effects depends on both the change of duty cycle over the length of the data stream as well as the time constant for the particular system and finally on the length of the data stream, so it is not possible to give a general histogram for this effect. Most of the time it is seen as a constant offset for a given duty cycle, or as a slowly varying offset along a pattern.

8.3.9 Bounded Uncorrelated Jitter (BUJ)

This last category is basically a catch-all for remaining jitter that does not fit into any of above categories. It is also labeled “deterministic” because it is always bounded and thus non-Gaussian and characterized by its peak-to-peak value.[91] Depending on whom one asks, “uncorrelated” is meant to stand either for “uncorrelated to known causes” or “uncorrelated to the signal under test”.

One common source is crosstalk from adjacent data lines. Of course it will make a difference if the data signal on these lines has some correlation to the data signal on the path under test (in which case in may possibly look like data dependent jitter) or not. Or it could be power supply noise

[91] Unbounded components would get lumped into random jitter.

(switching power supplies are especially bad offenders), or electromagnetic interference from some non-periodic source.

Due to the large variety of causes it is not possible to make too many general statements about the resulting jitter distribution. But since it is bounded it has to fall off to zero faster than a Gaussian distribution outside some limit.

9. JITTER ANALYSIS

9.1 More Ways to Visualize Jitter

We have already seen that jitter distributions can be characterized by their timing histogram (i.e. the jitter PDF). But depending on the way the signal is acquired as well as the preferred jitter analysis method there are several other ways to display jitter.

9.1.1 Bit Error Rate

When dealing with digital data transmission, the emphasis from a system point of view is less on the actual signal but rather the data (expressed as digital bits) that gets transmitted. Naturally then the figure of merit when dealing with jitter is how many of the bits are transmitted correctly, and how many are corrupted (i.e. incorrect). This leads to the definition of the so-called "bit error rate" (BER) as

$$BER = \frac{\text{number of bit errors}}{\text{total number of bits transmitted}}. \tag{76}$$

For our purposes we are only interested in bit errors caused by jitter (i.e. we disregard errors that are due to functional failures of the transmitter, that is, where the logical bit stream contains an error. The mechanism here is that a transition to a logical state occurs either too late (so the receiver strobes the previous bit) or too early (so it strobes the next bit value instead of the current – expected – one), the displacement being caused by jitter[92]. Figure 65 plots a sequence of two transitions (two bits), overlaid with the timing jitter distributions for each of the transitions (assumed to be equal), as

[92] We should note that for correct (or incorrect) data bit transmission it is of no importance if the jitter affects the transmitted signal itself or the sampling strobe of the receiver (usually both will exhibit jitter). It is the *relative* jitter between signal and strobe that counts. The total jitter of the transmission is the convolution of signal jitter and strobe jitter.

well as the sampling instant (assumed to be somewhere between the ideal (average) transitions points. The probability of a bit error is the sum of two probabilities: First, the probability that the first transition arrives too late, and second, the probability that the second transition arrives too early. This total probability is denoted by the shaded portion under the curves in Figure 65.

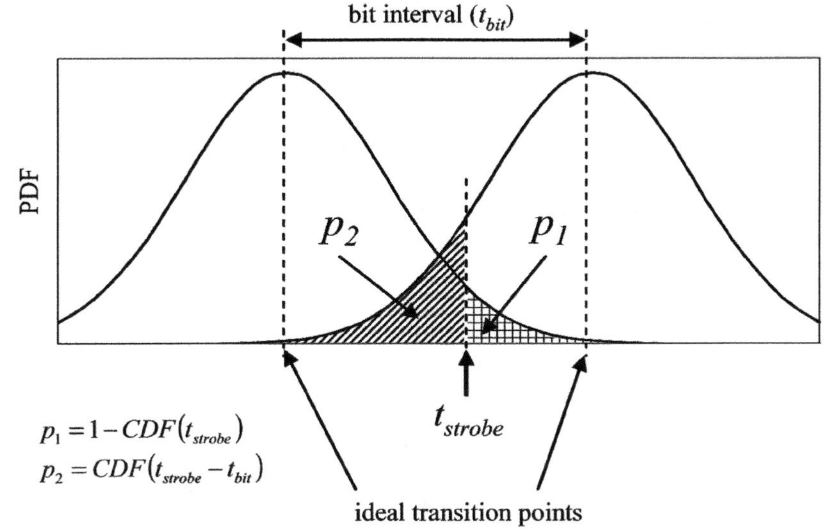

Figure 65: The bit error rate is given by the sum of two probabilities – the first transition happening too late, or the second happening too early.

In addition, to get the bit error rate from this distribution, we need to know how many transitions actually occur – unless we are dealing with a simple clock signal, there will always be sequences of two or more bits of the same value (e.g. ones) one after another. In other words, the number of transitions will always be a fraction of the number of transmitted bits. The ratio between number of transitions and number of bits is called "transition density". For a clock this density would be unity (i.e. 100%), while for typical more or less random data streams it is often around 0.5 (i.e. 50%)[93]. A transition density lower than 100% will reduce the bit error rate by the same factor – if there is no level change between one bit and the next (because both are high or both are low), then there is no possibility for a bit error due to jitter since their values are equal and it makes no difference if the receiver strobes in one or the other. The bit error rate is thus

$$BER = \text{error probability} \times \text{transition density}. \tag{77}$$

[93] A transition density of 0.5 is usually assumed unless explicitly stated otherwise. This is a good approximation e.g. for (pseudo-)random data streams, or data that uses 8b/10b encoding, but there are many practically important cases where the transitions can be much more parse (e.g. memory patterns or other algorithmic patterns).

As an example, let's assume we have a data stream with a jitter distribution that is purely random (Gaussian), just as displayed in fig. 65. For this special distribution we already know the probability $p(n)$ for the edge falling outside some interval of $\pm n\sigma$ (refer to Table 2 in section 6.4.1). Furthermore, let's assume the bit interval is $2n\sigma$ long and the strobe is centered in the interval, so it sits at $n\sigma$ distance from the two ideal edge positions.

The probability for the first transition to occur too late is then $p(n)/2$, and for the second transition to occur too early it is also $p(n)/2$, adding up to a total probability of $p(n)$. Assuming a transition probability of 0.5 the BER becomes $0.5 \times p(n)$. BERs for different values of n are given in Table 3. For serial data transmission schemes the required performance is commonly a BER of less than about 10^{-12}, which according to this table translates to a necessary minimum margin of somewhat over $\pm 7\sigma$ for purely Gaussian jitter distributions (assuming the strobe is centered in the bit interval).

Table 3: Bit error rates for a Gaussian distribution, for different confidence intervals, assuming a transition density of 0.5, and the strobe centered in the bit interval.

interval	BER
$\pm 1\,\sigma$	0.159
$\pm 2\,\sigma$	2.28×10^{-2}
$\pm 3\,\sigma$	1.35×10^{-3}
$\pm 4\,\sigma$	1.84×10^{-5}
$\pm 5\,\sigma$	2.14×10^{-7}
$\pm 6\,\sigma$	9.87×10^{-10}
$\pm 7\,\sigma$	**1.28×10^{-12}**
$\pm 8\,\sigma$	6.22×10^{-16}
$\pm 9\,\sigma$	1.13×10^{-19}
$\pm 10\,\sigma$	7.67×10^{-24}

9.1.2 Bathtub Plots

In the previous example we placed the sample strobe in the center of the bit interval, which seemed the most logical thing to do since the distribution was symmetrical and centered on the ideal position.[94] A very valid question is now what would happen if we moved the strobe away from the center towards one of the two ends of the interval. This is not merely a theoretical consideration, because in any real system the strobe may very easily be offset from its ideal position due to receiver circuit timing tolerances, skew in the transmission path, jitter on the strobe generator, and so on. Moving the

[94] And indeed in this case this is the position that yields the lowest BER.

strobe closer to one of the (ideal) transition points will reduce our margin for signal jitter and consequently deteriorate the BER, and it is important to know what our margin for such strobe displacements is before the BER becomes too large.

If we move the strobe all the way to one end of the bit period, the error probability will be very close to 0.5 if the distribution is reasonably centered around the ideal edge position: Let's assume we moved it to the left end (i.e., early). If the first transition occurs only the slightest bit late, there will be a bit error, and since the transition is assumed to be centered on that point, the probability for this to occur is 50%[95]. On the other hand, since we are rather far away from the next transition, the probability of this transition already occurring is negligible compared to 50%. (If that weren't the case, meaning the width of the jitter is large against the bit interval, the system BER would be so high as to make the system unusable in the first place). If we move the strobe now further and further towards the middle, the observed BER will drop all the way until we reach the center of the bit period, and after that will increase again because we are approaching the next transition point.[96] The BER is the failure probability multiplied by the transition density, so the BER will actually rise to 0.25 at both ends of the bit interval if we assume a transition density of 0.5.

We can plot this behavior – BER versus the strobe position – over the whole bit period[97]. Since the range over which the BER can vary usually spans many orders of magnitude, it is useful to display the BER on a logarithmic scale.[98] The resulting graph is called a "bathtub plot" since (with some imagination) it often resembles the silhouette of a nicely shaped bathtub. Figure 66 displays an example for such a bathtub plot. From this plot it is fairly easy to estimate the margin one has in the strobe placement in order to hit a specified BER.

[95] provided there *is* a transition at this particular bit boundary, the probability of which is given by the transition density.

[96] The distribution shown here is symmetrical around the bit boundary. In practice this is not necessarily the case. First, deterministic jitter can have highly asymmetrical distribution. Second, while in electrical systems thermal noise usually gives the same value for σ on the left and the right side of the bit interval, in optical systems nonlinearities in the laser transmitter can cause deviations from this behavior.

[97] Another common name for "bit period" is "unit interval"(UI).

[98] As stated, the BER at the edges of the interval will be close to half the transition density, while at the center it may be 10^{-12} or less for real-world transmission schemes.

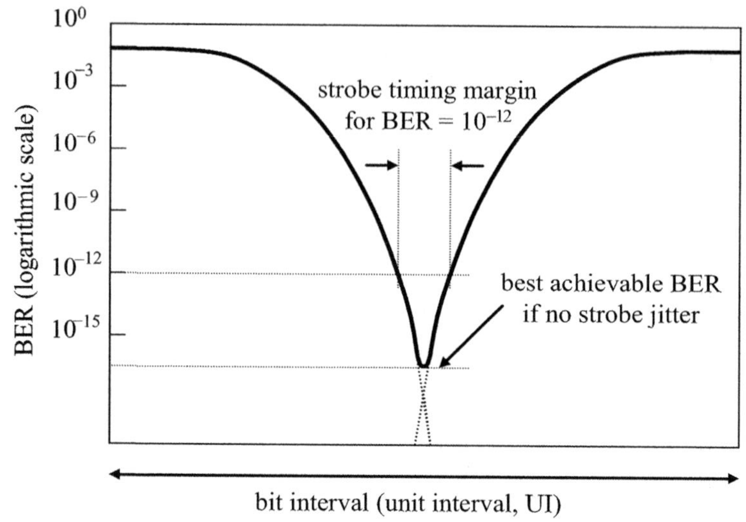

Figure 66: Bathtub plots are a common way to visualize and extrapolate the bit error rate for different strobe positions.

As we have already stated, the BER is the sum of two probabilities, namely the probability p_1 that the first transition occurs later than the strobe, and the probability p_2 that the second transition occurs already earlier than the strobe, multiplied by the transition density D. Looking back at Figure 65, we can write this down mathematically as

$$
\begin{aligned}
&p_1 = 1 - CDF(t_{strobe}) \\
&p_1 = CDF(t_{strobe} - t_{bit}) \\
&\Rightarrow BER = D \times \{1 - CDF(t_{strobe}) + CDF(t_{strobe} - t_{bit})\}
\end{aligned}
\qquad (78)
$$

where t_{strobe} is the position of the strobe in the bit period (zero being the left end of the bit interval), and t_{bit} is the length of the bit interval. From this we see that there is a very close relation between the jitter CDF (and hence the jitter PDF) on one hand and the bathtub plot on the other.

9.1.3 Eye Diagrams and How to Read Them

The last way to visualize jitter is the so-called eye diagram. While jitter histograms and bathtub plots rely on edge positions only (i.e. on the timing value where the signal crosses a certain amplitude threshold), eye diagrams are a more general tool because they give information about both the timing and the amplitude behavior of the signal.[99] Since they require information in both the voltage as well as the timing direction, they are a visualization tool mostly found when working with oscilloscopes, but both modern BERT instruments as well as digital testers employing level comparators are able to generate them, too.

An eye diagram consists of a superposition of many (or all) bit periods of a signal, aligned in a way so the nominal edge locations and voltage levels are aligned. This process is illustrated in Figure 67. Since it does not matter which bit periods are used – as long as they are a representative sample of the total bit stream – we can even use an equivalent-time sampling scope that only captures a single point of each interval.

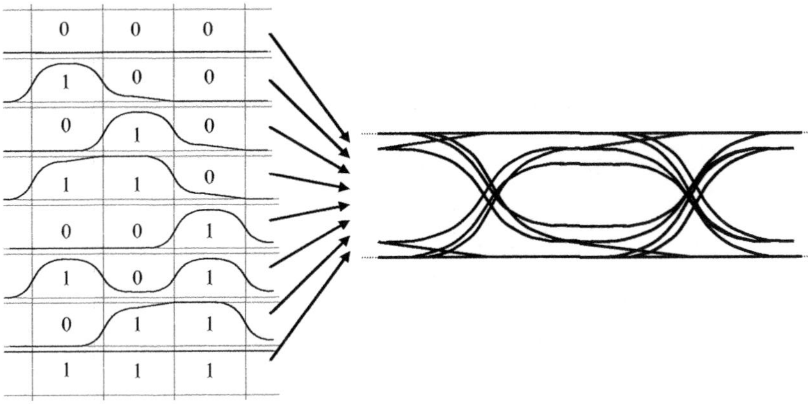

Figure 67: Relationship between a data bit stream and its eye diagram.

The bit periods used may be adjacent ones, common when the eye diagram is generated through post-processing based on a long continuous bit stream signal captured by a real-time sampling scope. In this case the nominal positions are usually inferred from a least-squares fit of the average bit period. These may or may not be the actual position of the true reference clock, which can lead to jitter results that do not match the actual system performance.[100] Or the periods may be isolated periods (or even single data points, see before) randomly taken from somewhere from the data stream – which is the case when the eye diagram is acquired directly by triggering on a separate clock signal that runs at the bit rate of the data stream. In this case the nominal positions are given by the clock edges, so if the clock has some

[99] While in this book we are mostly interested in timing measurements, as we have seen there is often a close connection between timing errors and voltage noise (or other level aberrations), so having information about the voltage domain can still be very valuable even for pure timing considerations.

[100] Remember the discussion of the ideal timing position.

jitter, this will add directly to the visible jitter in the eye diagram. The clock itself can again be some external reference signal, or the clock recovered from the date stream itself (by either a hardware clock recover circuit or in software) for the case of an embedded clock.[101]

On a scope the most common method to display the eye diagram is to run it in infinite persistence mode. Since there will soon be a large number of traces on the screen that will be impossible to keep apart – after all, the screen resolution is limited – color-grading or grayscales are used to indicate how often a particular point (pixel) on the screen has been hit by a trace.[102]

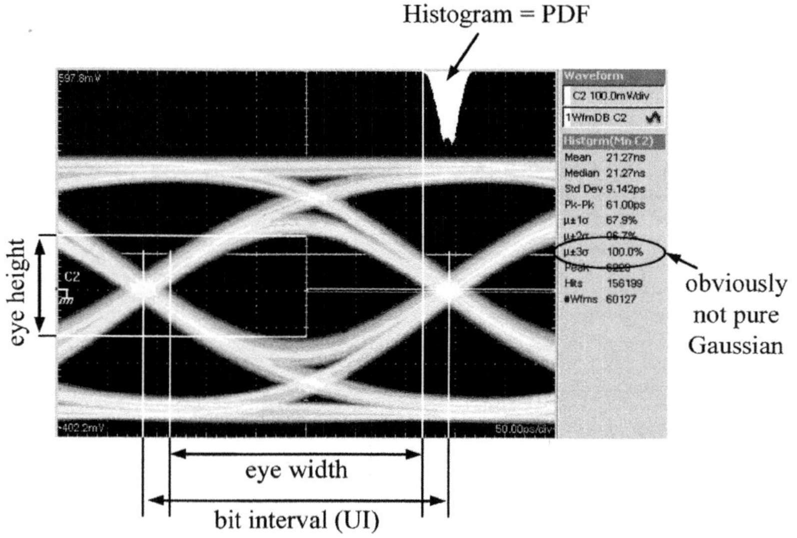

Figure 68: Example of an eye diagram using grayscale coding. It also shows a few parameters that can be extracted from the eye diagram.

Figure 68 shows such a display (in grayscale mode). With a little imagination it looks like an eye, which explains where the name comes from. There is a large amount of information obtainable from the eye diagram. The software running on modern oscilloscopes is usually able to provide best fits for rise and fall times, static high and low level, bit period,

[101] We should strongly note that it is not a valid approach to trigger on the signal under test itself, especially if it is not a clock waveform but rather a data stream: This will only yield the cycle (or N-cycle) jitter, and we already know that this is very different from the true signal-to-clock jitter!

[102] Displaying in grayscale mode makes the display very similar to what one would get on an analog scope where the persistence is provided by the afterglow of the electron tube and the persistence of vision of the human eye.

and peak-to-peak as well as RMS jitter.[103] In addition there is a close relationship between the eye diagram and the jitter PDF: The PDF can be obtained by simply setting up a horizontal histogram at the voltage threshold, as is also done in Figure 68.

The horizontal span at the center of the diagram at the threshold level where there are no hits is called the "horizontal eye opening", and gives the range of strobe positions that would result in zero BER (for the given number of acquisitions, that is). Consequently the vertical span of the "signal free area" at the strobe position is called the "vertical eye opening". Eye diagrams make it also fairly easy to spot duty cycle distortion – in this case the rising band of traces would be offset relative to the falling band, and their crossing point would be offset from the nominal threshold.

In summary an eye diagram gives an easy and intuitive overview over general jitter (and noise) behavior of a signal. On the other hand, since it lumps the whole pattern into a single picture, one loses the exact sequence of jitter values, data bits, and waveforms. Thus it is a good visual tool to roughly judge jitter and waveform fidelity quickly (and a tool that has been around since the times of analog scopes), but not too great for profound analysis. Color grading is no match for the precise numerical information that e.g. a bathtub plot yields. For numerical jitter analysis having the whole data stream is preferable, provided this is available (only real-time sampling scopes are fully able to provide this).

9.1.4 Eye Diagrams vs. BER vs. Waveform Capture

The final question to ask in this chapter is – which method of displaying jitter is appropriate for which situation? An ideal eye diagram would show the composite of all possible events (bit sequences), no matter how infrequently they occur. But in practice, a typical eye diagram consists of only relatively few signal traces, maybe a few 1000, especially when acquired with an instrument with a low effective sample rate like an equivalent-time sampling scope. This means that while the eye diagram gives a good *overall* picture of the signal performance, it will most likely miss any rare events – but those events will still cause the system to fail in its application!

[103] Since – except maybe for very clean clock signals that would not give a true eye diagram anyway – there are virtually always both random and deterministic jitter components present, the value of both peak-to-peak and RMS jitter is of limited value: The former will change with the number of samples taken (since the random component is unbounded), while the latter is not sufficient to fully describe the jitter in presence of deterministic (non-Gaussian) components.

A first solution to this problem is to capture the full waveform over an extended interval and construct the eye diagram through post-processing the data. A real-time sampling oscilloscope is the tool of choice here. The advantage is that it will actually capture every cycle within the capture time frame, but even the most advanced (and expensive) scope have memory depths of less than 100 million samples and thus limit the captured data stream to maybe 10^7 bit intervals or less. One could repeat the acquisition and merge all the separate eyes together, but processing time quickly renders further improvement prohibitively time-consuming, so one is unlikely to achieve measured coverage beyond a BER of maybe 10^{-9}, and most likely much less.

One systematic weakness of eye diagrams is that they only show *parametric* information of the waveform, e.g. overshoot, rise times, or jitter. They will give no indication if the system has a logical problem – i.e. if at some instant it was meant to send a 1, but instead sent a 0 – the 0 will look nice and healthy in the diagram, but will make the system fail in its application.

Finally, a BERT box can sample data at the actual data rate without interruption for arbitrarily long bit streams if it uses pattern looping or an algorithmically generated pattern. That means it will capture each and every failure in the bit stream, even the rarest ones, and both logical ones and ones that are caused by excessive jitter or insufficient signal integrity. At data rates in the Gigabit-per-second range BER coverage of 10^{-12} and beyond is possible with measurement times of a few minutes. On the other hand a single such scan gives little information as to why and how an error occurred – for such investigation a real-time oscilloscope is more useful.

However it is not sufficient to just run e.g. 2×10^{12} data bits (assuming a transition density of 0.5) without a single error to guarantee a BER of 10^{-12}. Such a BER only says that *on average* there will be a single error in a data stream of this length. The actual outcome is statistical and random (since those errors are “rare events” the distribution is given by the so-called Poisson distribution which for large number of “positive” events (here, errors) approaches a Gaussian distribution. As a result, some stretches of 2×10^{12} data bits will contain no errors at all, while others will contain more than a single bit error. For our measurements this means that we have to observe at least several bit errors (meaning we have to run several times 2×10^{12} data bits for our example) to be able to guarantee our measured BER with a sufficient confidence.

The Poisson distribution for our case is

$$P(a) = e^{-m} \times \frac{m^a}{a!}, \tag{79}$$

where m is the true average number of bit errors in the data stream, a is the actually measured number of bit errors, and $P(a)$ is the probability to find a bit errors in the interval. For example, measuring a data stream of 10^{12} bits with a true BER of 10^{-12} will on average produce 1 bit error ($m = 1$). If we perform the measurement and indeed find none ($a = 0$) or just one ($a = 1$) error, the confidence that the BER is truly no more than 10^{-12} is

$$P(a = 0 \text{ or } a = 1) = e^{-1} \times \frac{1^0}{1} + e^{-1} \times \frac{1^1}{1} \approx 0.74\,, \qquad (80)$$

i.e. our confidence is just 74%.

9.2 Jitter Extraction and Separation

9.2.1 Why Analyze Jitter?

In the early days of jitter analysis – just a few years in the past actually – all that most engineers had available to measure jitter were analog scopes. With such an instrument the best one can hope for is to eyeball the peak-to-peak jitter from the washed-out trace or eye diagram on the screen. Modern digital instruments on the other side can provide much more information on the exact timing distribution, trend over time, etc.

As we have seen, different types of jitter have different characteristics as well as different causes. As an example for the former, while random jitter is unbounded and one needs statistical, probabilistic treatment to be able to predict long-term error rates, deterministic jitter types can often be characterized by hard peak-to-peak numbers. As to the causes, periodic jitter is often due to crosstalk (e.g. from a power supply), data dependent effects are often caused by reflections or path losses, and so on. And in virtually any real system we always have a mixture of several different jitter types. So if we manage to separate and quantify the different contributors, it can enable us to perform two important tasks:

1. We can predict (extrapolate) the long-term behavior of our system under test from observation (measurement) over only a limited time span. With "behavior" we mean things like peak-to-peak jitter, expected bit error rate under various conditions, timing margins, and so on. This allows for example to assert the correct function of a system over time spans of days or years from measurements that may take only a few seconds.

2. We can use knowledge of the jitter contributors to hunt down (and hopefully eliminate) their causes, and so improve the system performance. For example, if we determine that we have periodic jitter of a certain frequency, we may find that this frequency corresponds to the switch rate of our power supply. Or we may notice that the dominant contributor is data dependent jitter, prompting us to improve our transmission path.

Digital real-time sampling scopes are the easiest tools to use for jitter extraction because they give us the full data stream (i.e. the position of every single edge) and all we need to do (a bit simplified) is take the edge positions and put them into whatever statistics tool we would like. Their main limitation is the maximum signal speed (data rate) they can work with. All other instruments (time stampers, equivalent-time sampling scopes, bit error rate testers) give us only some form of aggregate, convoluted jitter result that we (or some software) must try to pick apart into its components. e.g. all we may have is some probability density function in the form of a histogram from an equivalent-time sampling scope, or a bathtub plot from a BERT box. This "picking apart" then always needs certain assumptions (since we have no way of knowing the actual behavior beforehand, e.g. correlation between jitter components) that may or may not be completely valid. That said, there are good reasons to use those types of instruments nevertheless, be it because they can go to extremely high bandwidths, or because they can perform the acquisition much faster than other instruments. In the following sections we will therefore assume that all we have is the probability density histogram of our data stream. This type of representation even makes sense in the case of a real-time oscilloscope (where in principle we can have all the raw data for millions of edge positions and no need to muddle them together in a single histogram), because unlike computers humans have difficulty grasping large sets of numbers, but can visually analyze two-dimensional plots with relative ease.

If we want to do some serious, accurate jitter analysis, the best choice is usually to take recourse to professional-grade software to crunch the data. But many engineers feel rather uncomfortable using such a tool in a "black-box manner" – sure, it always gives us *some* set of numbers as result, but how do we know those numbers are correct? Thus we will now look for simple signatures and manual methods and rules of thumb to spot different types of jitter and to quantify their size. Once we can do that, we can use those methods to gauge the trustworthiness of our jitter software. And often all that is needed in the first place is an order-of-magnitude approximation – e.g. if we can say with confidence that the deterministic jitter is negligible in comparison with random jitter, we may not need to put in any more effort to improve our signal path, but rather concentrate on improving the driver noise of our system.

9.2.2 Composite Jitter Distributions

Before we go ahead and talk about how to *separate* jitter components present on a data stream, it is worthwhile to go for a moment into the opposite direction and have a look at how different types of jitter *add up* to create the total jitter distribution.

If two different jitter components are independent from (uncorrelated to) each other, then the total distribution is the convolution of the two.

If both distributions are discrete (e.g. duty cycle distortion and data dependent jitter), then the resulting histogram looks as shown in Figure 69: Each of the N components of the first distribution is split into a little "sub-histogram" of M components looking like the second distribution, and the total histogram now consists of $N \times M$ components.

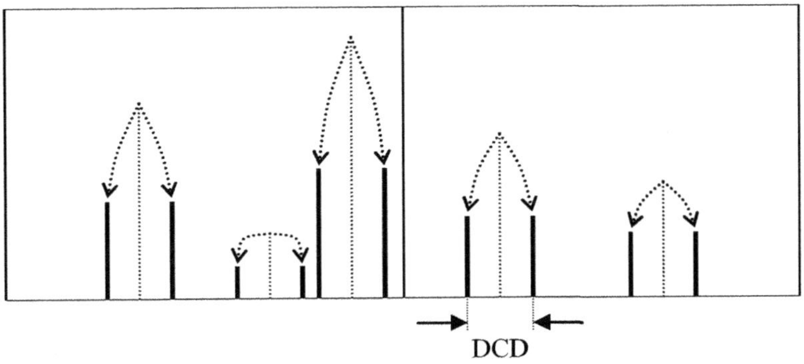

Figure 69: The combination (convolution) of two discrete jitter distributions – here, PDJ and DCD – results in another discrete distribution.

If one distribution is discrete (with N components) and the other continuous, than this resulting total histogram is a set of N continuous distributions added together. This was displayed before in Figure 64(b) for the case of data dependent jitter (discrete) in combination with random jitter (continuous Gaussian peaks). If the continuous distribution is wider than the separation between the discrete components, then the peaks run into each other and can get difficult to distinguish from each other (given that any practical measurement always has some amount of noise and uncertainty). Overall, one could say that the discrete distribution gets "smeared out" by the continuous one.

Finally, if both distributions are continuous, then again a smearing occurs (but this time it is a matter of personal taste to decide which distribution smears out which). As an example, we can look back to Figure 60(b) to see a combination of periodic jitter and random jitter.

9.2.2.1 Combining Random Jitter Components

Fortunately for us, truly random jitter components are usually uncorrelated, so the convolution assumption holds true. Moreover, given their known statistics and behavior (random events with Gaussian distribution), mathematics tells us that they add geometrically (at least as long as all components in the transmission path are behaving in a linear fashion), i.e. if we have N random jitter components with standard deviations σ_1, σ_2, … , σ_N, the total random jitter will be

$$\sigma_{total} = \sqrt{\sigma_1^2 + \sigma_2^2 + \ldots + \sigma_N^2}\ . \tag{81}$$

9.2.2.2 Combining Deterministic Jitter Components

As we know deterministic jitter is characterized by a peak-to-peak spread rather than a standard deviation. The peak-to-peak spread of the convolution of *uncorrelated* distributions is the sum of the peak-to-peak spreads of all the contributors, so people often assume that the peak-to-peak deterministic jitter (DJ) is the sum of all deterministic jitter components. Unfortunately (or fortunately, depending on one's point of view) this is not always true because – unlike for random jitter – we have no assurance that the deterministic jitter sources are truly uncorrelated: For example, if we send a signal with pattern dependent jitter through a transmission channel, the response of the channel (timing location of reflections, and so forth) will be different than it would were there no DDJ on the signal. Correlation between jitter components can – if anything – only reduce the total jitter spread because it may prevent the worst case of component 1 to occur at the same time as the worst case of component 2, thus the total peak-to-peak spread DJ_{total} is given as

$$DJ_{total} \leq DJ_1 + DJ_2 + \ldots + DJ_N\ , \tag{82}$$

where DJ_1, DJ_2, … , DJ_N are the peak-to-peak spreads of the individual jitter contributors.

9.2.2.3 Combining Random and Deterministic Jitter

As should have become obvious by now, separating the different jitter contributors based on a composite histogram can become quite daunting very easily. This is why we will spend the rest of the chapter looking at how this can be achieved in practice. But before we do that, it is a good moment to debunk a very common myth about the combined effect of random and deterministic jitter.

The hallmark of random jitter is that it is unbounded, and it is thus characterized by its standard deviation σ. There is no upper theoretical limit on the possible maximum timing error caused by random jitter, but since larger deviations become rapidly less probable we can always give some limit $\pm n_{rand} \times \sigma$ (i.e. an interval of $2 \times n_{rand} \times \sigma$) outside which the BER is smaller than some limit (e.g. around 10^{-12} for $\pm 7\sigma$ assuming a transition density of 0.5).

On the other hand the hallmark of deterministic jitter, independent of its exact cause, is that it is bounded, so we can specify a peak-to-peak value Δx_{det} that it will never exceed – in other words, the BER for strobe instants outside this interval is exactly zero.

If both random and deterministic components are present at the same time we are obviously again stuck with an unbounded total distribution, so like for pure random jitter we have to specify the total jitter width in combination with the residual BER. A very common assumption (or better: misconception) is that the resulting interval is $\Delta x_{\det} + 2 \times n_{rand} \times \sigma$. This definition can even be found in official communication interface standards, which nevertheless does not make it correct. To illustrate the point, let's consider a signal whose jitter consists of a simple bimodal distribution (two peaks of equal height separated by Δx_{det}, e.g. because of duty cycle distortion) and some random jitter of standard deviation σ. For the sake of simplicity let's use $n_{rand} = 1$, which for the random jitter alone would mean approx. 32% of the edge population is outside the interval, resulting in a BER of about 16%[104]. In Figure 70(a) we see that those 32% are represented by the area under the Gaussian bell curve outside $\pm 1\sigma$.

Let's look what happens if we put random and deterministic jitter together: Now we have *two* Gaussian peaks whose centers are separated by

[104] This small interval is chosen because it makes the subsequent graphical illustration easier. In practice one would of course use a smaller BER, commonly somewhere between 10^{-3} and 10^{-12} (corresponding to n_{rand} between approx. 3 and 7). All the conclusions in the text stay valid no matter what confidence interval we chose. We remember that the BER is given by the population in the fringe lobes outside the confidence interval, multiplied by the average transition density. We assume a density of 0.5 throughout this book unless stated otherwise.

Δx_{det}, as shown in Figure 70(b). The data stream still has the same number of transitions, so the total area of each peak is only half of the height of the single peak. Since the widths (given by the standard deviation) are the same as before, this means the height of each of the two peaks is only half the height of the single peak in Figure 70(a). The total distribution is simply the sum of the two peaks. For the sake of clarity in the subsequent discussion we assume that Δx_{det} is of the order of or larger than σ.

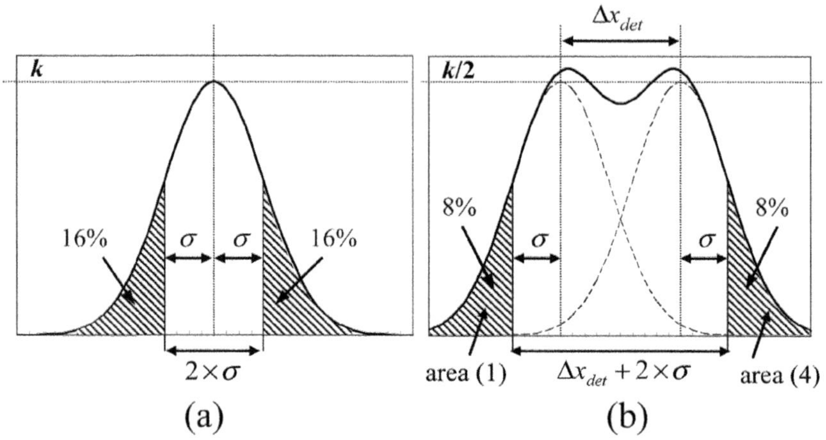

Figure 70: Confidence intervals of $\pm 1\sigma$ for (a) a single Gaussian peak, and (b) for a bimodal distribution consisting of two Gaussian peaks. Note that even though they were plotted equally high to conserve readability in (b), the true single-peak height (*k/2*) in (b) is only half of the height *k* in (a) so the total area stays the same.

Figure 70(b) also shows the interval $\Delta x_{\text{det}} + 2 \times n_{rand} \times \sigma$. Now let's look at what our BER for this interval is, again given by the area under the total distribution outside the interval, multiplied by the transition density. This "error area" consists of four components: (1) The area of peak 1 left of the interval, (2) the area of peak 1 right of the interval, (3) the area of peak 2 left of the interval, (4) the area of peak 2 right of the interval. Since the peaks are Gaussian, they fall of rapidly with increasing distance from the center. That means that only two of the four components ((1) and (4)) contribute measurably to the error area, while the other two ((2) and (3)) have negligible impact. Each area contains one quarter of the original error area for the purely random jitter distribution (which was approx. 32% of the peak area), namely approx. 8%. So the total error area is just 16%, for a BER of only 8%.

In other words, the crude approximation formula overestimates BER and total jitter! Using it will cause an overly conservative jitter estimate, which may trigger difficult, time-consuming and costly over-engineering of the system to reduce the jitter, or failing the specification, while the system in reality performs much better than one assumes.

How much the formula overestimates jitter depends on a variety of factors: For simple bimodal deterministic distributions as the one considered in the previous example the real BER will be about 50% lower than the estimate if Δx_{det} is larger than σ. Things get even worse with multimodal or continuous distributions of sufficient width ($\Delta x_{det} \geq \sigma$): As an example we may picture the same scenario as before but for a trimodal of three equally spaced, equally sized peaks instead of a bimodal distribution: Only two out of six partial areas contribute considerably to the BER, so the over-estimation of the BER is approx. 67%!

However, after all the criticism we should state that there is in fact a very good reason (other than ignorance) why so many people believe in the formula: Again as a direct result of the rapid fall-off of the Gaussian distribution with increasing distance (and thus rapid increase with decreasing distance), the BER increases rapidly when the confidence interval (given by n_{rand}) is tightened. E.g. going from $n_{rand} = 7$ to $n_{rand} = 6$ increases the BER from around 10^{-12} to 10^{-9}, a thousand-fold increase!

This means that even an initial overestimate of the BER by 66% can be corrected by a very small adjustment to n_{rand}, often so small that it is not even worth considering. Commercial jitter analysis software will do the necessary correction automatically. So after reading this section it should no longer be a reason for surprise (or a cause to call the vendor's customer support to report a software bug) when our jitter analysis software gives us numbers for the random and the deterministic component that do not exactly add up to the number for the total jitter width!

9.2.3 Spotting Deterministic Components

Probably one of the first things we want to know when looking at a brand-new signal is if there is any sizeable deterministic jitter contribution of any kind (can be data dependent, periodic, etc.).

The easiest case comes when we have large data dependent effects (or other deterministic effects that produce a finite number of discrete peaks in the distribution) that dominate over any random effects. Such a histogram is shown in Figure 71. Each peak is broadened by random jitter. With very good approximation we can determine the random jitter as the RMS width of a single peak, and the deterministic jitter as the peak-to-peak spread of the peaks' mean values. This is also indicated in Figure 71.

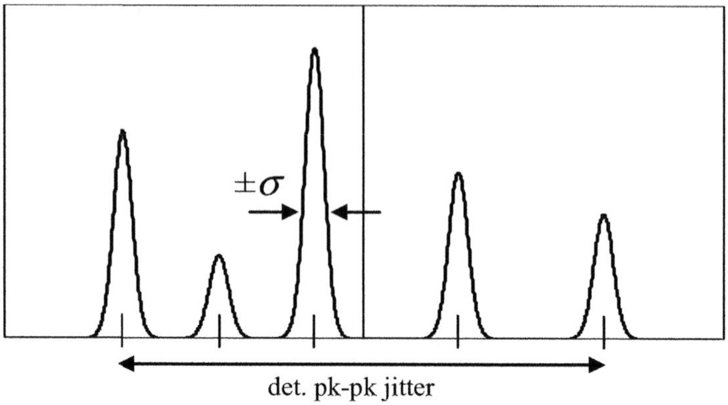

Figure 71: Combination of data dependent jitter and random jitter where the former is dominant.

Unfortunately such easy cases are rather rare in practice. Usually random jitter and deterministic jitter are either of the same order of magnitude, or (especially for clock signals) random jitter dominates, in any case making it difficult to even spot separate peaks in the broad Gaussian background. So we need slightly more subtle means to extract that information.

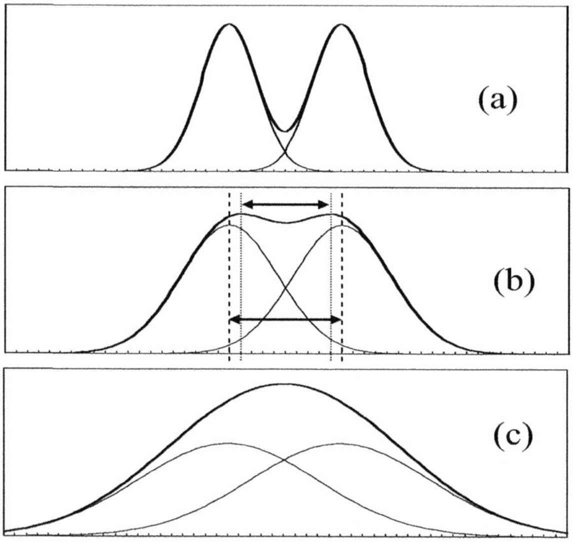

Figure 72: Combination of deterministic and random jitter for different amounts of the latter. The deterministic peaks smear into each other (a) and for sufficient random jitter (c) are no longer separately discernable. On the way to larger and larger random jitter the apparent peak positions deviate more and more from the true peak positions (b).

First let's assume we can still see peaks, but the random noise is large enough so they overlap considerably. As an example, refer to Figure 72 where we assume we have only two peaks (caused by deterministic jitter) of equal height and separated by 10 ps. With random jitter of only 2.5 ps RMS (Figure 72(a)) the two peaks are clearly separated and the distance of their maxima is close to the "true" separation of 10 ps. With larger random jitter of 4.3 ps (Figure 72(b)) the two peaks are broadened so much that they don't disappear at the respective other maximum; this results in a reduced apparent distance of 9 ps, a 10% measurement error. In other words, in general the peaks in the jitter histogram are *not* the exact locations of the deterministic components (though they may be close).

Figure 73 shows a graph that relates the error between apparent and true peak separation to the measured distribution width and the apparent peak separation, assuming a symmetric bimodal distribution (i.e. only two Gaussian components of same height and width make up the distribution). We see that as long as the standard deviation of the peaks is less than about a third of the peak separation, the apparent peak locations virtually coincide with the true peak centers. On the other hand, for standard deviations larger than the peak separation divided by $\sqrt{2}$ the composite distribution exhibits only a single peak. We can use Figure 73 to make our results a bit more accurate, but don't expect miracles!

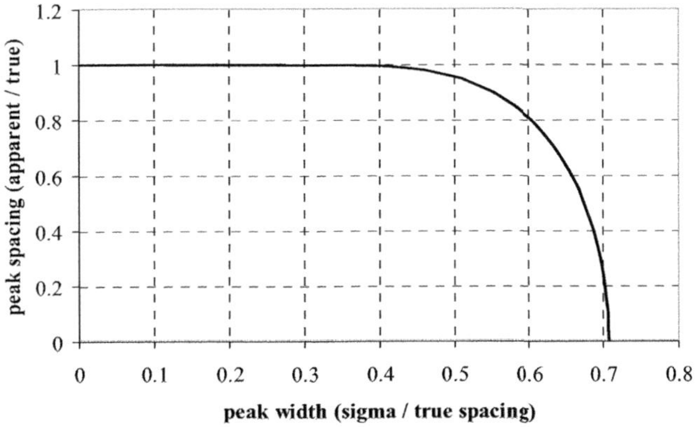

Figure 73: True peak separation vs. apparent peak separation for two Gaussian peaks of equal height and equal width.

If random jitter is so dominant that no separate peaks are discernable (Figure 72(c)) then we can hardly hope to extract too much information about the deterministic component without the help of some sophisticated curve-fitting program. But to see if there is any deterministic jitter at all one criterion we can use is that random jitter has a Gaussian distribution, while deterministic jitter usually does not – even though it may approximate Gaussian shape for not too large offsets from the mean value. We already know that random jitter – being Gaussian – is unbounded, while deterministic jitter is always bounded, i.e. its distribution vanishes for sufficient distance from the distribution center. Thus if we just go out far enough, a deterministic distribution (or a mixture of random and deterministic components) will always fall off faster to zero than a pure Gaussian distribution: The former will contain a larger portion of its area between $\pm k\sigma$ than the latter if k is sufficiently large.

A good way to check this is to take the calculated standard mean, area, and standard deviation of the measured histogram and use those numbers to plot a pure Gaussian curve with the same parameters. When overlaid with the measured histogram it is easy to spot systematic deviations. The bulk of the measured curve will be higher and wider (because eventually falls off faster), and it may even have a plateau rather than a peak in the middle (but again fall off faster on the fringes). This test is illustrated in Figure 74.

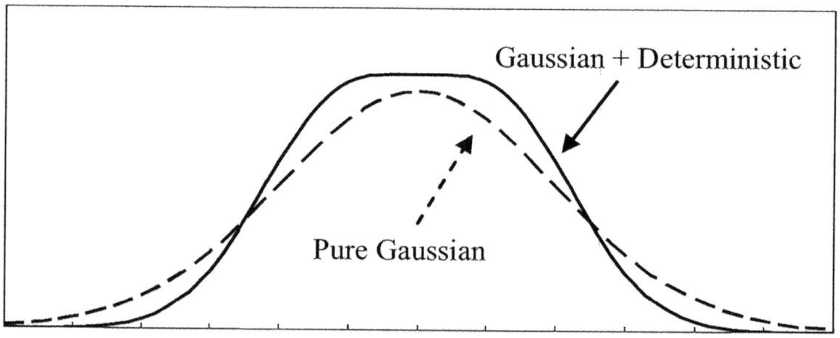

Figure 74: Pure Gaussian distribution compared to a distribution with same mean value and standard deviation.

For a "quick and dirty" check we can do even easier: Many oscilloscopes will tell us directly not only σ but also the percentage of hits between $\pm\sigma$, $\pm 2\sigma$, and $\pm 3\sigma$. From Table 2 we know that those percentages should be 68.3%, 95.4%, and 99.7%, respectively, if the distribution is Gaussian. If those numbers deviate considerable from those we measure, then we can be sure to have some deterministic components in our distribution. However we need to pay attention that the number of data points (hits) making up our histogram is large enough to allow for statistically significant results. A percentage of 99.7% means we need about 1000 hits just to get 3 outliers, which is a small number likely to fluctuate wildly when we repeat the acquisition (remember that the variation is approximately proportional to $1/\sqrt{N}$, and $N = 3$ in this case). In order to get reliable results, we should wait to accumulate at least about 100000 hits. The number of hits is important if we compare them to a Gaussian curve like described before, and we should choose about the same number lest we are guaranteed to end up with some jagged measurement curve that won't resemble anything useful.

9.2.4 A Word on Test Pattern Generation

When performing jitter measurements we can distinguish two general situations: First, we may have the possibility (or be forced to, depending on one's point of view) to measure the actual signal when the system under test is running its normal operation. What we measure in this case is the jitter of the system for this specific pattern, nothing more and nothing less. It does not tell us how the system would perform under different conditions and how bad the jitter could possibly get.

Second, during system development it is often not possible to generate exactly the pattern the system may be running later on – be it because it involves complicated encoding we are lacking the hardware for, or because (like e.g. a test instrument) the exact pattern will depend on which application the system will be put later into, and so on. Thus what we will be striving for are either "typical" or "worst case" numbers to get a feeling for the range of inaccuracies we will encounter later on.

Now how do we choose such "typical" or "worst case" pattern to exercise our system? Obviously we want a good match between the statistics of this bit stream pattern and the "real" application pattern. Important parameters here are average duty cycle (affecting duty cycle effects), longest stream of continuous ones and zeroes (affecting data dependent jitter), and the selection of sub-patterns contained in the pattern (again affecting data dependent jitter). For example, if the system is designed to produce an arbitrary data stream, testing it with a simple clock pattern will yield very poor coverage since it won't show any data dependent timing errors.

One of the most-used types of test data streams are pseudo-random bit streams (PRBS). "Pseudo-random" means the bit sequence has the same statistics as a truly random one, but it is actually generated algorithmically and thus is predictable and repeatable. Repeatability is a big advantage since it allows us to reproduce the circuit behavior again and again. The circuitry used to produce PRBS patterns are so-called "linear-feedback shift registers" (LFSR). LFSRs are extremely simple and are often included as test mode circuitry in signal source devices, and every modern BERT box offers built-in PRBS generation.

An LFSR is a shift register whose output and one or more additional taps are combined in an exclusive-or gate and the result fed back into the input of the shift register. If those taps are chosen correctly, such an LFSR consisting of N stages (flip-flops) can produce a repeating pattern of length $2^N - 1$ bits that contains *all possible* sub-patterns of length $N-1$, plus one sequence of

N ones.[105] As an example, Figure 75 shows such an LFSR for $N = 3$ as well as the resulting bit stream. A PRBS generated this way shows a small imbalance: First, the number of ones is $N/2$ while the number of zeros is $(N/2-1)$, so the bit stream is not exactly DC balanced (important when employing AC coupling in the signal path). Second, the longest continuous stream of ones is N bits long, while the longest continuous stream of zeros is only N-1 bits long. Fortunately the impact of these imbalances disappears quickly for larger values of N so for practical purposes ($N \geq 7$) a PRBS can be assumed to be balanced and symmetric.

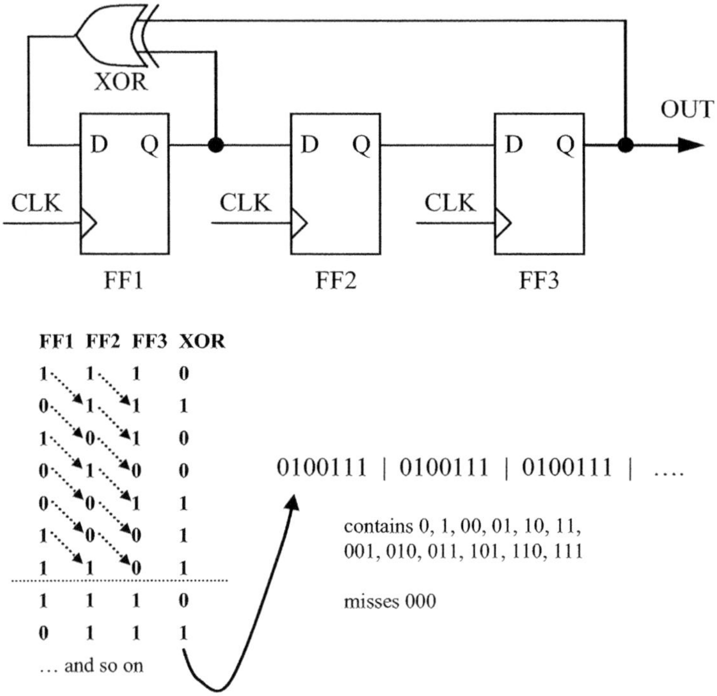

Figure 75: A linear-feedback shift register (LFSR) with proper feedback setup (taps) can create random-looking sequences that contain all possible bit sequences up to a certain length.

Such a bit stream is a pretty good match for many practical patterns. For example, let's take the case of a signal path that will later run 8b/10b encoded data, which is DC balanced and where the longest continuous sequence of either ones or zeros is five. This encoding scheme is heavily used in modern high-speed serial transmission schemes. A $2^7 - 1$ PRBS, i.e. a pattern length of 127 bits, in which the longest bit stream of ones and zeros

[105] This will only work if the LFSR is initialized with *any* state other than all zeros (in which it would get stuck and produce only zeros at the output).

is 7 and 6, respectively, is a popular choice as a test pattern since it gives a small error margin over the "real" pattern and it is reasonably DC balanced (imbalance 1/127 < 1%). On the other hand, a $2^{23} - 1$ PRBS is commonly used to test SONET or SDH telecommunication systems which require test patterns with lower frequency content (longer run lengths) to better approximate scrambled or random data.

A last remark pertains to the actual measurement of the pattern (either the "true" pattern or some pattern like the PRBS described before): While it is nice to have a pattern that *produces* all possible sub-patterns that the system can possibly encounter and this way will exhibit the full jitter possible, this does not help us at all if we don't actually *acquire and measure* all those sub-patterns. This may sound very logical, but in fact is a common trap to fall prey to. Let's assume we have a $2^N - 1$ PRBS pattern. If we have a real-time sampling scope or a BERT box and N is not too large, we can do a single-shot acquisition that spans at least $2^N - 1$ bit periods and are guaranteed to have captured every single event at least once. But for many other instruments (an equivalent-time sampling scope triggered on the clock signal, or a time interval analyzer) that sample the data stream asynchronously it may be a long time until we have finally hit every single bit period in the repeating pattern at least once. Even with a real-time oscilloscope we may run out of capture memory for very long data streams so we would have to revert back to triggering on the clock signal and acquiring an eye diagram.

Things are even worse with data streams that are not necessarily deterministic (at least in the sense that we as the test engineer can't predict after which time every single possible bit sequence has been produced, as we can very easily e.g. for a $2^N - 1$ PRBS). Some event may be very rare but nevertheless cause worst-case jitter. So in a nutshell, just because we have a long and complicated test pattern does not necessarily mean our test results are representative. If we don't measure long enough, a $2^{23} - 1$ PRBS may not give much better coverage than a $2^7 - 1$ PRBS.

What's more, in order to account for random jitter, acquiring each transition only once is still not enough: There is a good possibility (actually 50%) that when we hit the sub-pattern that causes worst case data dependent timing error on the subsequent edge, the random jitter on this particular edge reduces the total error and thus makes the total jitter distribution look narrower than it really is. One solution to this is to repeat the test patterns and thus to acquire each edge *many* times so the random jitter is virtually guaranteed to show up in either direction. Unfortunately this would usually increase the acquisition time to almost infinitely large numbers to give even a moderate level of confidence. The second, much more practical method is to determine each jitter contributor separately and then put them together mathematically. In other words, this is again a perfect reason why jitter extraction and jitter modeling make a lot of sense.

9.2.5 Random Jitter Extraction

One of the numerous myths about jitter is that "random jitter is the jitter one sees on a clock signal". By now it should be very clear to us that even a humble clock signal can suffer from a variety of jitter contributions – random jitter, duty cycle distortion, periodic jitter, and uncorrelated deterministic jitter. The only one it lacks is data dependent jitter.

So if generating a clock signal is not sufficient to isolate random jitter, how can we determine the size of random jitter on our signal? It depends on the instrument we have available, but there are some common strategies. Let's assume all we can acquire is the jitter histogram (i.e. the PDF of the total jitter). The first step is to look at the same edge over and over again[106], i.e. acquire the histogram for this particular edge. This way we are at least guaranteed to get rid of all data dependent timing errors.

The second step is to remember that random jitter is Gaussian and unbounded, while deterministic jitter is bounded, i.e. it is always restricted to some limited area. For simplicity let's assume the deterministic components consist of discrete peaks. Each of these peaks is of course smeared out by random jitter. Then, if we just go out far enough, we will always end up with significant contribution only coming from the outermost peak. The proof is easy: Assume two peaks, located (centered) at positions T and $(T\text{-}\Delta T)$, as shown in Figure 76. Each of them is Gaussian in shape with the same σ, and they add up to

$$PDF_{tot}(t) = \frac{1}{\sigma\sqrt{2\pi}} \times \left[\frac{1}{2} \times e^{-\frac{(t-T)^2}{2\sigma^2}} + \frac{1}{2} \times e^{-\frac{(t-(T-\Delta T))^2}{2\sigma^2}} \right]. \qquad (83)$$

The factor ½ is necessary to normalize the total area to unity. The relative contribution of the two peaks at some timing position t is then

$$\frac{e^{-\frac{(t-(T-\Delta T))^2}{2\sigma^2}}}{e^{-\frac{(t-T)^2}{2\sigma^2}}} = e^{-\frac{(t-(T-\Delta T))^2}{2\sigma^2} + \frac{(t-T)^2}{2\sigma^2}} = e^{\frac{-2t\Delta T + 2T\Delta T - \Delta T^2}{2\sigma^2}} \xrightarrow[t \to +\infty]{} 0. \qquad (84)$$

[106] Same meaning an edge preceded by the same pattern every time. For a clock signal this can be any edge it the signal, for a data signal we will most likely choose to have the pattern repeat over and over again and to look at the n-th edge into the data stream.

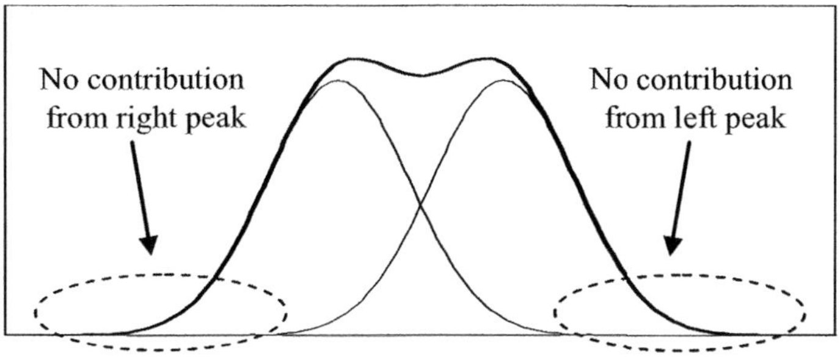

Figure 76: Distribution consisting of two overlapping Gaussian peaks: On the left and right fringes the peak closer to the respective side is always dominant.

In other words, if we go out far enough to the right, only the rightmost peak (centered at T) contributes, so the "tail" on the right fringe of the distribution will be almost purely Gaussian. The same consideration – for the other, leftmost peak – applies to the left side.

So if we continue assume that the random (Gaussian) component is the same for each peak, we can fit each of the fringes of the distribution with a Gaussian distribution which will give us the standard deviation σ of the random jitter.[107]

Of course if there are no other jitter components (other than random jitter and data dependent jitter), then looking at a single edge indeed gives a purely Gaussian histogram (but this always needs to be checked!) and the random jitter extraction becomes almost trivial.

A very different method is to extract the random jitter by looking at the data stream in the frequency domain, usually employing a high-performance spectrum analyzer. Random jitter in this case shows up as side lobes around the main spectral lines. This can be a very sensitive method and we will discuss it in detail in section 10.3, when we look more closely into special spectrum analyzers techniques.

Some solutions on real-time sampling scopes or time interval analyzers employ a mixture of time and frequency domain methods; they acquire the waveform (or directly the sequency of periods for a TIA), calculate the jitter trend and Fourier-transform it[108]. Random jitter then shows up as noise floor in the Fourier spectrum, and the total noise power (integrated over the

[107] We should note that this "tail fitting" algorithm is copyrighted by Wavecrest Inc., so one can't use it in any commercial product, but it is a good introduction into jitter analysis methods. Jitter analysis software from other manufacturers uses different strategies to obtain the random jitter contribution.

[108] Note that Fourier-transforming jitter numbers (i.e. a sequence of transition timings) is not the same as Fourier-transforming the original signal itself!

frequency range of interest) can be related to the standard deviation of the random jitter.

Finally a word of caution if we attempt to measure random jitter on an equivalent-time sampling scope by acquiring the edge position repeatedly. Do *not* use the automated crossing point measurement and look at its statistics (standard deviation) – this is *not* the true jitter of the data signal! The reason is that in order to calculate a threshold crossing, the scope needs to use at least two data points (the one just before and the one just after the crossing). But since a scope of this type does not do single-shot acquisitions of the waveform, these two points come from two completely independent acquisitions, separated in time by the rather low actual sampling rate. The crossing point is now based on the average between those two points, and as we know the standard deviation of an average over 2 samples is smaller (by a factor of $\sqrt{2}$) than the standard deviation of the single samples. In other words, this method would underestimate the random jitter by roughly 30%. The correct way to perform the measurement is to acquire a histogram at the threshold level and look at its standard deviation.[109] Comparing the results of the two methods in a real setup is a very worthwhile and instructive exercise.

9.2.6 Periodic Jitter Extraction

Just as random jitter, periodic jitter can be tough to pinpoint since it is easiest to see with real-time acquisition. If we are in possession of the full stream of edge positions from a single-shot acquisition (e.g. by using a real-time sampling scope), our task is relatively simple: All we have to do is a Fourier transformation of the jitter trend (not of the signal itself!), and the periodic components will then stick out as peaks in the spectrum. The peak positions give the frequency of each jitter component. The size of each peak (more precisely: its area) is a direct measure for the size of the particular jitter component.

As second real-time method uses the frequency domain representation of the signal (this time it is the signal, not the jitter trend!), i.e. the signal spectrum is acquired with a spectrum analyzer.[110] Periodic jitter (and jitter in general!) can be regarded as a form of phase modulation, and since the spectral distribution of a phase modulated signal is well understood it can be

[109] Because real-time sampling scopes acquire the full waveform in a single shot, they don't have this issue, and either measurement technique (crossing point statistics or histogram, respectively) will yield the correct result.

[110] Theoretically a real-time sampling oscilloscope could be used and the signal Fourier-transformed (which high-end scopes can do automatically), but chances are high the resolution and especially the dynamic range of the spectrum will not be sufficient for this type of analysis.

used to extract the size and the frequency of the periodic jitter. We will dig into the details of this in section 10.3.1.

What's more, it turns out that as long as there is sufficient random jitter on the signal, even humble digital testers and BERT boxes can extract periodic jitter because it will result in a periodic variation of the BER along the bit pattern. We'll outline the method in detail in the next chapter as well (section 10.2.5).

If all we have it the jitter histogram (i.e. the jitter PDF), then things are more complicated since this does not give any information about the sequence in which the single data points were acquired. The best we can strive for is to extract the size of the jitter, but the jitter frequency is irretrievably lost. If the periodic jitter is the dominant contributor then determining its amplitude is relatively easy – the peak-to-peak jitter is approximately the peak-to-peak width of the histogram. If random jitter is not negligible, one can first determine its size (through tail fit or some other method, as described in the previous section) and then mathematically remove it from the distribution.[111] Then – provided no other jitter components are present – we can fit the remaining distribution with the periodic jitter PDF presented in section 8.3.5.[112]

9.2.7 Duty Cycle Distortion Extraction

As far as the data acquisition is concerned, things will get easier now. While random and periodic jitter are best analyzed based on a full single-shot (real-time) acquisition of the data stream (since they usually vary from one signal burst to the next when the signal is repeated), duty cycle distortion, data dependent jitter, and duty cycle (thermal) effects are deterministic in the sense that they will yield the same result every time the measurement is repeated if the measurement conditions remain unchanged. This means if necessary we can acquire the data set a little bit at a time, like e.g. an equivalent-time sampling scope or a time interval analyzer has to do it anyway.

Since duty cycle distortion is a timing offset of rising edges vs. falling edges, the most straightforward way to determine its size is to look at those two groups of edges separately, determine their average positions (from their PDFs) and then take the difference between the two means. This method implicitly assumes that all other jitter contributors affect the two groups identically, which for most – but not all! – data streams is a good

[111] The necessary mathematical tool is the so-called *deconvolution*, a process described at the end of this book (in section 11.3).

[112] And if the shape of this PDF does *not* fit well onto our measured distribution this will be a clear sign that other types of jitter are indeed present.

assumption.[113] Most of the instruments we discussed are capable of distinguishing between edge polarities.

Even in eye diagrams it is relatively straightforward to see duty cycle distortion: Unless the eye is largely closed, one can distinguish a band of rising edges and a band of falling edges, crossing each other. For each of them one can determine the center line and where it crosses the threshold.[114] The timing difference in the crossing points is the duty cycle distortion.

If all we have is the jitter PDF of the data stream as a whole then the situation is not as easy. Like for periodic jitter we first have to get rid of the random jitter to clear up the picture. After that two well separated, identical peaks should be left (while periodic jitter would have caused a distribution that is continuous between the two peaks, as in Figure 60). Of course if other jitter components are present the histogram will look more complicated and most likely will not give enough accurate information for reliable jitter component extraction.

9.2.8 Data Dependent Jitter Extraction

By definition data dependent jitter is the part of jitter that makes the position of a specific edge dependent on the "history" on the channel, i.e. which data pattern preceded it. The extraction method of choice depends on the situation and on how much access one has to the choice of pattern driven (i.e. if we have to characterize the data dependent jitter on a given data stream, or if we can vary the pattern to our liking).

One of the most straightforward methods to extract data dependent jitter is to compare the jitter histograms of the data stream with the jitter histogram of a clock signal coming from the same driver. Since clock signals don't have any data dependent jitter[115], the difference between the two resulting histograms is a direct measure of the effect of data dependent jitter. It can

[113] For example, in a stream of bits like ...00100100100100.... the rising edges are always preceded by a different pattern than the falling edges, so they will experience a different data dependent error that in this signal stream is indistinguishable from "true" duty cycle distortion.

[114] How this threshold is defined depends on the particular situation. For a differential signal for example it is almost always the zero crossing, i.e. an absolute signal level. But in some cases we may need to infer the threshold from the measured signal itself – in this case we have to determine high and low level from the top and bottom band of traces and then use e.g. the mean between those two values. High-end scopes usually have those analysis functions already built in.

[115] At least except a constant timing offset. Another way to see this is that each edge is preceded by the same pattern and thus suffers the same, constant data dependent error. But this will not affect deconvolution except again for a constant timing offset of the result.

then be extracted by deconvolution[116] of the pattern histogram with the clock histogram.

Since data dependent jitter is deterministic, i.e. it will repeat exactly when the measurement is repeated under unchanged conditions, we can employ averaging to remove random, periodic and uncorrelated deterministic jitter, which greatly simplifies our task (not just for the previous method, but for any method that only shoots for data dependent jitter extraction). Often it isn't even necessary to perform any deconvolution since after averaging there may not be many other jitter contributors left. It also means we can use virtually any of the instruments described earlier, with the exception of spectrum analyzers which are a poor fit or this type of jitter.

A second method would be to generate all possible N-bit long patterns and determine the influence of each on the immediately following transition. Averaging is again highly recommended to weed out other jitter contributions. If the pattern cannot be changed at will, but provided it is sufficiently long and contains a variety of different sub-patterns we can do basically the same analysis. Pseudo-random bit streams are an excellent choice because they are guaranteed to contain every possible pattern up to some length N, and especially in serial transmission setups they are often available as a built-in test mode of the transmitter.

One important consideration when using the results obtained with any of those methods to predict jitter under real-world conditions is the actual population of each data dependent jitter component. e.g. the worst case pattern may never occur in the real application, so our peak-to-peak jitter estimate will be too conservative if it did include this worst case. On the other hand the minimum and the maximum worst case pattern could be the most frequent, so our distribution would be heavily concentrated on the fringes, which would not affect the peak-to-peak jitter, but would still increase the BER because it increases the number of edges that have large timing errors.

A special case are linear systems (e.g. we are looking only at the data dependent jitter added by the – passive – transmission path, or our signal source is operating entirely in its linear range[117]). Here we can in principle gather all the information needed to calculate data dependent jitter for arbitrary patterns from the measurement of a single isolated transition (e.g. low to high) – the so-called step response of the system:

[116] We will discuss deconvolution towards the end of this book. In a nutshell in the present case it is the mathematically reversial of the smearing of the pure data dependent jitter histogram by the other jitter sources.

[117] This usually implies that the rising and falling transitions have the same shape (only inverted) regarding rise time, overshoot, etc.

Any arbitrary bit stream can then be thought of as a superposition of step responses, each shifted in time to match the corresponding edge position, and inverted in polarity for falling edges. As we know, for a linear system the response to a sum of inputs is the sum of the responses for each isolated input, which in turn is just the step response we measured. This way we can predict the resulting waveform for any digital pattern we choose, and from that determine the threshold crossing points and the timing jitter. An example is shown in Figure 77.

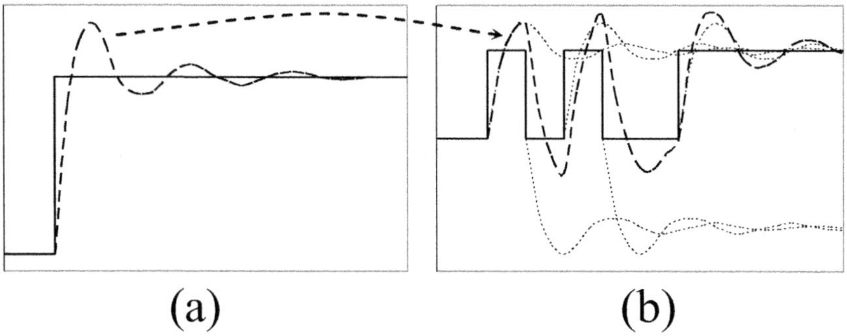

Figure 77: For perfectly linear systems one can predict the response to an arbitrary bit pattern based on the measurement of a single transition.

Another way to see this is that each transition causes some signal aberrations (overshoot, ringing, settling) that linger on the line for some while. At each edge location all the residual aberrations from previous transitions simply add up (after all it's a linear system), and we know that signal level aberrations (i.e. noise) inevitably cause proportional timing aberrations (i.e., jitter).

9.2.9 Extraction of Duty Cycle Effects

As discussed, duty cycle effects are caused by thermal changes depending on the average duty cycle of the signal. Thus the basic strategy seems very clear and easy – keep one edge (the "victim") unchanged at the same programmed place and vary the duty cycle of the rest of the pattern. Figure 78 shows schematically how this could be implemented. We can again employ averaging to improve the clarity of the effect.

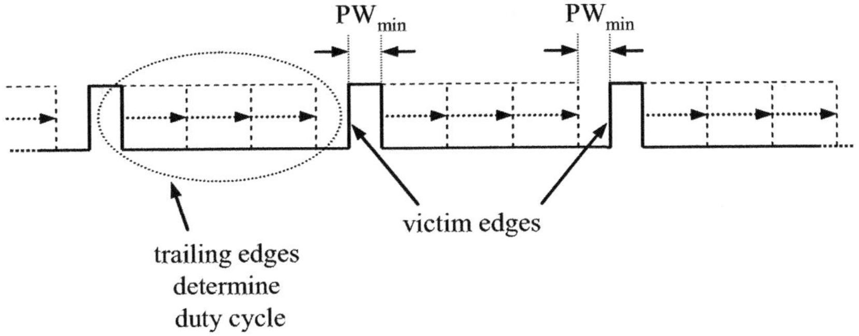

Figure 78: Extraction of duty cycle effects: The leading edge of a high-going (or low-going) pulse is moved to vary the duty cycle, while the trailing edge is observed.

The trouble comes with the details. Since it is always best to look at each effect in isolation we don't want to mix duty cycle effects with other jitter components. Averaging gets rid of most, but not of data dependent errors. Fortunately the latter are relatively short term, i.e. usually disappear after times longer than just a few bit periods[118], while thermal effects take much longer time scales to take hold, usually in the order of hundreds of nanoseconds or even tens of microseconds. So the solution is to keep the signal static before our victim edge for some time longer than the time scale of the data dependent effects. Any timing change that we see now when we change the duty cycle will be caused entirely by duty cycle effects.

However we should not become overzealous and make the distance of the rest of the edges in the pattern excessively large. This could lead to the case when the die temperature settles out partially (or completely) between transitions, so changes in the duty cycle would only result in a reduced effect (or none at all) no matter what the duty cycle is, and as a consequence we would underestimate the size of the effect. Some experimentation may be required to find the right tradeoff.

Finally, again because the effect is settling rather slowly, we need to make sure we give the device (or system) under test enough time after the start of the pattern to settle into thermal equilibrium before we attempt our measurement. In a benchtop setup this is usually not an issue since the time scale is usually well below a millisecond, and test time is well above a second, but it could be a source of systematic measurement error on automated test systems – the results would become dependent on the length of the warm-up period (or the lack thereof).

[118] And after having extracted the data dependent jitter we should have a pretty good idea how many bit periods its effects keep lingering around.

9.2.10 Uncorrelated Deterministic Jitter

There really is no generally valid concept to analyze this jitter. As for jitter extraction, uncorrelated deterministic jitter is any jitter that is left over when all the previous jitter types are accounted for. This may or may not be a large portion of the total jitter. Depending on the system one has to make an educated guess as to the possible sources and then vary the suspected cause. For example if we suspect crosstalk from adjacent signal lines, we would vary the data pattern on those lines (or keep them completely silent), see if this changes the observed jitter and if yes, by how much. Or if external electromagnetic fields are a possibility we may want to shield our system better, decouple its power supply from the main supply through a transformer, etc., and see if that changes the uncorrelated jitter component.

9.2.11 RJ/DJ Separation and the Dual-Dirac Model

If we have control over the data and timing sent, and if we can measure the edge timings directly (e.g. with a real-time scope), it is usually possible to extract the different jitter contributors with good confidence. But unfortunately in many cases we won't be so lucky, and all we have is a convoluted jitter PDF or a bathtub plot. There is no generally valid way to separate all the single jitter components from such a plot with absolute confidence; this is compounded by the fact that our sample size (the number of transitions acquired) is always limited, and there is always some noise (uncertainty) on any measurement, meaning that it is often difficult to decide if a small feature in the PDF is really caused by a separate jitter component, or if it is just due to our limited measurement accuracy. An example would be densely spaced data dependent jitter peaks smeared out by a wide background of random jitter – section 9.2.3 illustrated that it is then very difficult to separate those deterministic peaks from each other. Second, the Central Limit Theorem in statistics states that if we convolute a large number of independent (uncorrelated) influences together, e.g. several different jitter types, the final distribution will closely resemble a Gaussian distribution, even though all contributors may be deterministic and the distribution is in fact still bounded.

On the other hand, what is really fundamentally important to us is to know the long-term performance (given by the BER) of our system. The most important issue is here the relative contributions of deterministic jitter (DJ) – bounded and thus never exceeding a certain limit, the peak-to-peak spread – and random jitter (RJ) – unbounded (characterized by its standard deviation σ) and thus potentially fatal to our data integrity if we just wait for sufficiently long times.

This is where the so-called dual-Dirac model comes in: It *assumes* that the total jitter distribution is simply the combination of two Gaussian peaks whose mean values are separated by a distance $DJ_{\delta\delta}$. Such a distribution can also be seen as the convolution of a bimodal deterministic distribution (two Dirac delta functions separated by $DJ_{\delta\delta}$) with a Gaussian distribution of width σ, hence the name "Dual-Dirac Model". This is exactly the same distribution as the one produced by duty cycle distortion plus random jitter (see Figure 62(b) in section 8.3.6, and Figure 72 in section 9.2.3).

We cannot emphasize enough that the use of this specific distribution is a pure assumption to enable easy jitter modeling for arbitrary measured distributions. With the exception of duty cycle distortion or some other square-wave like jitter trend the modeled peak separation $DJ_{\delta\delta}$ will *not* be equal to the true peak-to-peak width DJ_{true} of the deterministic jitter component. In most other cases fitting the measured jitter distribution with the dual-Dirac model will underestimate the deterministic component and overestimate the random jitter, i.e.

$$DJ_{\delta\delta} \le DJ_{true}, \quad \text{and } \sigma_{\delta\delta} \ge \sigma_{true}. \tag{85}$$

As a result the model will predict a higher-than-real BER and thus too small a data eye if used to extrapolate measured data to lower BERs, because far away from the bit boundaries it is the random, unbounded jitter that dominates the BER – and doing this sort of extrapolation is often the main reason for modeling jitter in the first place. An example for this behavior is Figure 60 where the deterministic contribution is caused by periodic jitter. While the true peak-to-peak periodic jitter is $2a$, a fit with the dual-Dirac model would place the two Gaussian peaks somewhere close to the peaks of the total distribution[119], i.e. at a smaller distance.

This is an important piece of information because many instruments – BERTs, TIAs, etc. – that offer jitter analysis capabilities base their RJ/DJ separation and their BER estimates on the dual-Dirac model. On the upside this makes for easy and relatively consistent jitter modeling across different instruments and platforms, and the results tend to be comparable to each other, so it provides observable, well defined quantities for use in technology standards. However we must not make the mistake and confuse the reported values for $DJ_{\delta\delta}$ and $\sigma_{\delta\delta}$ with the true peak-to-peak deterministic jitter DJ_{true} and the true random jitter width σ_{true}, and we should be aware that BERs predicted based on those values will be overly conservative. The mismatch is dependent on the specific distribution and especially on the ratio between

[119] actually even a bit closer because the central range contains significant area and thus a lot of weight in the curve fit.

σ_{true} and DJ_{true}. σ is typically overestimated by less than 10%, while the underestimate of *DJ* can be much more dramatic. Since random jitter dominates at small BERs, the extrapolated total jitter will usually also be too large by a few percent. If in doubt it always pays to confirm if the particular instrument or software used is relying on the dual-Dirac model or if it uses some more sophisticated modeling approach.

9.2.12 Commercial Jitter Analysis Software

As we have seen the visualization and especially the numerical analysis of jitter require a fair amount of mathematics. Fortunately it is not necessary to do those tedious manipulations by hand or develop customized software: Virtually all manufacturers of high-end oscilloscopes, time interval analyzers, or BERT boxes offer jitter analysis tools for their instruments. In addition there exists also third-party software that works with one or more of those instruments.

Let's consider the upside of these tools: First, since the vendor has put in a considerable amount of effort and manpower into the development of this software, today's jitter analysis programs tend to be fairly comprehensive, well debugged, and user-friendly. Often a basic jitter analysis and separation of different jitter types is the work of just a few minutes (provided, of course, that one is familiar with the user interface of the particular program). They provide many different visualization tools (like eye diagrams, bathtub plots, and jitter trend plots) that allow the user to get a good feeling for the signal timing quality. They are able to automatically dissect complicated jitter histograms and/or jitter trend plots and decompose jitter into its components (like random or data dependent jitter). Finally they can perform automated performance predictions (like BER or jitter extrapolation – more to that in the next section).

After so many good things, we need to mention that they have some downsides as well: First, they tend to be expensive, running into thousands of dollars for a single license. This is absolute overkill if one only needs occasional, order-of magnitude jitter estimates, or is dealing with rather slow signals (and large jitter).

Second, different tools greatly differ in the algorithms they use for jitter separation, one reason being that many methods (like the tail fit to get random jitter from a histogram) are under copyright by a single vendor, and others are forced to find different solutions. The specific modeling assumptions may also be questionable in certain situations, the dual-Dirac model being just one example. In addition, as we know different types of instruments have greatly different capabilities and methods to acquire the signal (e.g. a time interval analyzer can only capture a few edges at a time, while a real-time sampling scope can acquire a long waveform in a single shot). The result is that while each tool seems to give very precise and fine-grained jitter results, good correlation between different tools (and instruments) from different vendors – and even between different software packages and different instruments of the same vendor – is often difficult to achieve. And since for a user those software packages are basically a "black box" – feed in the waveform on one end and get out a set of jitter numbers on the other – he has virtually no possibility of deciding which result – if any! – is the correct one based on the software alone.

All this is not intended to give the impression those software tools are of little use. On the contrary, they can be a big help and save valuable time and effort *if used and understood properly* and if we always approach the results they yield with a good portion of skepticism. Therefore it is a big advantage to have good knowledge of the instruments' properties with regard to signal capture, and of the behavior of different jitter types and how to separate them – basically all that we covered so far. With that knowledge we can then go and try to understand and sanity-check the results. If we do have doubts, or if we encounter non-correlation between instruments, we can try to gain more knowledge by asking questions like:

- Do the two instruments really acquire the same type of data? Or does e.g. one acquire edge-to-edge jitter, whereas the other acquires edge-to-reference-clock jitter? What are the differences in acquisition mode?
- What is the ideal timing (the reference clock) derived from? An actual clock signal, or hardware clock recover, or software clock recovery, etc.? For clock recovery, what are the parameters (e.g. PLL bandwidth, PLL jitter)?
- Are the two methods equally sensitive to long-term variations? e.g. if a real-time scope acquires many repetitions of a shorter pattern in a single run, it will see long-term drift of the source's time base which could potentially be very large (in the actual system the receiver's PLL would follow those slow variations in the bit rate). On the other hand an equivalent-time sampling scope that gets triggered at each pattern repeat would only see the variation during one short pattern repeat, thus be blind to most of the drift and as a result could yield much "better" (i.e., smaller) jitter numbers.
- Is the analog performance of the instruments comparable, or has one instrument higher bandwidth or less intrinsic time base jitter than the other one? (Lower bandwidth would result in additional data dependent timing errors, but potentially less intrinsic noise). How do the acquired waveforms compare (e.g. one instrument may show more overshoot than the other one)?

- What type of data is the jitter analysis based on in each case, like histogram, jitter trend, full waveform, etc.?
- What assumptions or simplifications does the modeling software use, e.g. dual-Dirac model on one instrument vs. more elaborate decomposition on another?
- Would it be possible to isolate one or more of the jitter components with an independent method, preferably a "pedestrian" one (like the ones outlined in the previous sections) that produces easy-to-understand results and where we have full control over data acquisition and data processing? e.g. if the tools differ in the amount of periodic jitter they yield, we could measure it directly with a spectrum analyzer. Or for random jitter, instead of driving the full-blown data pattern, simplify it to a clock pattern and see if that changes the numbers and if it improves correlation. In this case we can even acquire a histogram ourselves and – provided it passes the test for pure Gaussian distribution – determine the standard deviation easily by hand.

9.2.13 Jitter Insertion

By now we have talked a lot about how to analyze and decompose jitter measured on a given signal. But in practice it is very often also necessary to generate a signal with well-known jitter characteristics that is then fed into the system under test, e.g. to test the jitter tolerance of a receiver, or the stability of a phase-locked loop.

The most straightforward approach is to start out with a clean signal, i.e. one that has negligible jitter, and then degrade this signal by adding some well-controlled amount of jitter of the desired type(s). Depending on the jitter type, the primary source for it can e.g. be an RF-generator (sine wave source) for periodic jitter, a white-noise source[120] for random jitter, or for maximum flexibility an arbitrary waveform generator (AWG). To generate data dependent timing errors, the most common solution is to insert a low-pass filter or a section of lossy transmission line (e.g. a long cable) into the transmission path. Of course the flexibility of such a solution is not too great since a change in the jitter characteristics requires a physical change to the path, e.g. inserting a cable of different length.

Figure 79 shows a possible setup that allows injection of random, periodic, and data dependent jitter into the signal generated by a pattern generator running off an external clock: A sine wave generator frequency-modulates (or phase modulates) the clock source driving the generator, so

[120] A well-controlled white noise source in terms of noise amplitude and noise bandwidth is e.g. the low-pass filtered output of a PRBS generator.

the pattern has already some periodic jitter. A low-pass filter at the generator output then adds data dependent jitter. A power combiner hooked up to a random noise source as well as to the output of the filter adds random noise that, as we know, translates into random jitter. Finally, a limiting amplifier (basically a digital transceiver) restores the levels so the final signal has only timing jitter, but very little level noise. This level restoration may or may not what the particular application (or communication standard) requires for testing, so any real setup may chose to use somewhat different approaches, e.g. omit this limiting amplifier. For example, as long as the peak-to-peak width of the inserted random jitter component is small against the rise time of the signal, the approach shown in Figure 79 for the random jitter would well, but it would break down for larger jitter values – then one would have to use the delay input of the generator (as done for the periodic jitter), but this may be limited in the achievable bandwidth. Or one could add any type of jitter by sending the data signal through an analog delay line – where the absolute delay depends on the signal level present on some control input – and driving this control input with the appropriate jitter signal.[121] For duty cycle distortion, adding a constant offset to the signal (e.g. by biasing it through a power combiner) would do the trick.

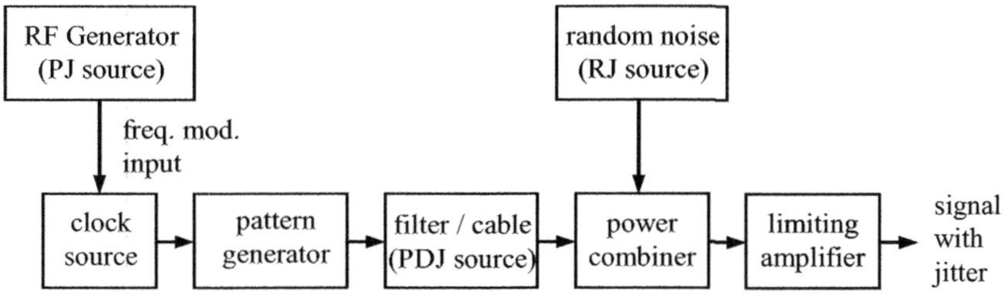

Figure 79: Generation of a bit stream with will defined, controllable jitter characteristics.

We should mention that jitter insertion through the clock source modulation input (as done for the periodic jitter in Figure 79) or through a variable delay line is nevertheless *not equivalent to* insertion of noise through the power combiner. The first adds a delay proportional to the jitter signal, which affects rising and falling edges the same (either both are early or both are late). Noise insertion on the other hand affects them in opposite directions: If the noise signal at a given time is positive, a rising edge will reach the receiver threshold earlier, i.e. the edge timing will be early; at the same time a falling edge will be late. Just picture a simple clock signal when

[121] If the required jitter bandwidth is not excessively high, a digitally controlled delay line driven by a digital data source (e.g. a microprocessor) would allow for a very flexible jitter generation.

inserting a relatively low-frequency sinusoidal jitter signal (jitter frequency much lower than the data rate): In the first case (delay input) it will cause low-frequency periodic jitter – first all the edges are early, then all become late, and so on – while in the second case (noise insertion) one edge will be early, the next one late, and so on, i.e. the jitter will contain extremely high frequency components of up to half the data rate. If the clock receiver drives a PLL, the first case will pose no problem, but the second could completely throw off the PLL – it won't be able to lock onto these fast changes.

9.3 Jitter Performance Prediction (Extrapolation)

Let's say we have successfully decomposed jitter on a signal into its components, using one of the methods described above. One purpose of this is often to know which types of jitter are hurting timing accuracy most and then go after their source. Once a fix is implemented, re-measuring (and decomposing) the jitter will show if it was successful in reducing the jitter.

A second, equally important application is the prediction of long-term performance (especially the BER) of a system based on a relatively short-term observation. The most fundamental task for this is the separation of random jitter (unbounded) from the deterministic contribution (which is bounded by some peak-to-peak value).

9.3.1 Extrapolation of Random Jitter

As we have seen more than once so far, random Gaussian jitter is fully characterized by a single number, namely its standard deviation σ (the second parameter, the mean value, is by definition the "real" edge position). Using Table 2 we can always find out how likely it is that we find events that fall outside $\pm n\sigma$, no matter what n is, in other words we can give a random jitter interval for a given BER even though we may not have even measured enough edges to find a single one of them that far away. That's the beauty of having a good theoretical knowledge of the jitter distribution.

In a bathtub plot random jitter is what determines the shape of curve when approaching the center (and thus the depths) of the "bathtub" (since deterministic jitter is bounded and will cease to play a role for distances from the bit interval's borders larger than its peak value). The width of the eye opening is then given by the distance of the left and the right intersection of the bathtub plot with the horizontal line at the specified BER, as shown in Figure 80 which indicates that the part of the curve that actually got measured does not extend to the required BER. It should be clear that such an extrapolation is the more reliable the closer the smallest measured BER is to the specified BER.

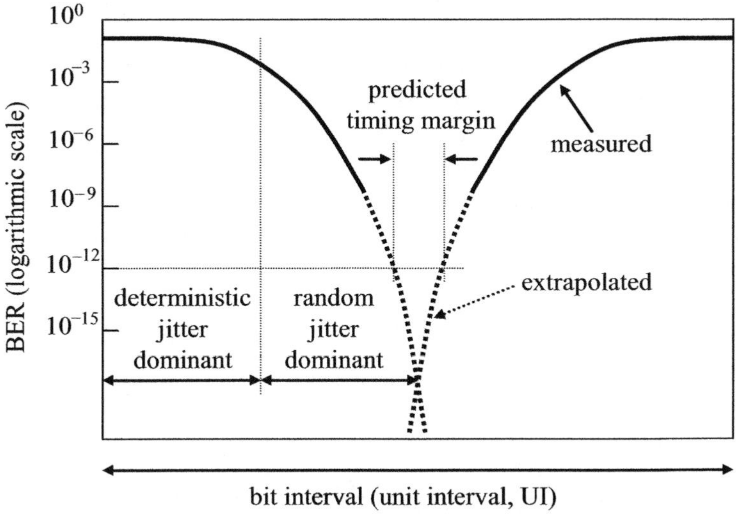

Figure 80: Bit error rate extrapolation from a bathtub plot.

Automated jitter analysis tools can do a pretty decent job at such *estimates*. Their advantage is that they give reasonably accurate results based on reasonably fast data acquisitions. However, they can't *prove* that the system's BER is really that good – the ultimate test required to prove the system fulfils its spec is always a test with a BERT box running a sufficiently long pattern – at least several times the inverse BER (1/BER) of bits.

9.3.2 Extrapolation of Deterministic Jitter

Deterministic jitter components are at the same time easier and more difficult to extrapolate than random jitter. Easier because we know that they are bounded and thus don't expect any further growth once we've established their peak-to-peak width. More difficult because other than random jitter (where the standard deviation is well established and stable after a relatively small number of samples, and we know well how to predict further growth in the peak-to-peak width at a given BER) it is sometimes an art to assure *when* we have captured the full breadth of events of some deterministic component. Especially for data dependent jitter we need to make sure we cycled through all possible patterns up to a reasonable length and actually also captured them (as explained earlier, a long pattern alone does not help if we don't acquire most or all of its transitions). Second, we need to make sure the pattern we measure is a good match to the pattern that will later be driven through the system in the "real world".

9.3.3 Prediction of Worst-Case Data Dependent Errors

A very neat theorem called "Maximal Linear System Response" allows predicting the absolute worst case data dependent jitter for a *linear, time invariant* system.[122] The basic idea here is to excite the system with a signal that causes maximum resonance.

The (somewhat simplified) strategy is outlined in Figure 81: The signal levels are assumed to be symmetrically centered around zero with a steady-state amplitude of $\pm A$ (a different DC offset is easy to take into account by re-normalizing the levels[123]). We will also assume that the rise time and the fall time (and the rising and falling wave shape) of the signal are approximately equal.

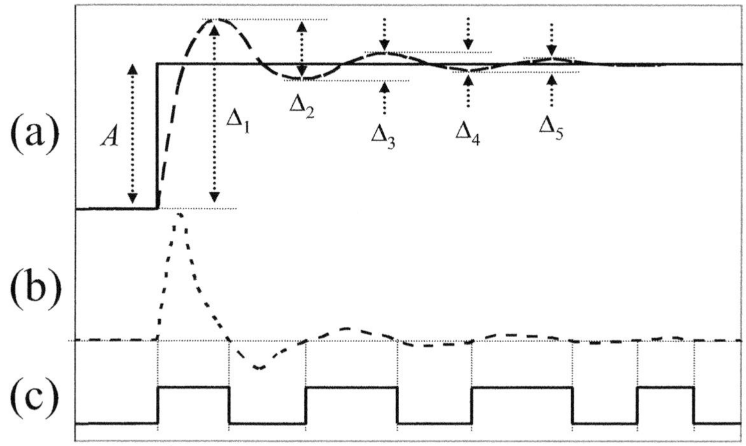

Figure 81: Constructing the worst-case stimulus pattern for a linear system. (a) Step response, (b) Impulse response (derivative of the step response), (c) Worst-case digital stimulus. Note that the intervals of the transitions are not multiples of some bit period, but can have arbitrary lengths, so they usually don't correspond to any real bit pattern.

The first thing to determine is the so-called step response of the system, i.e. the output waveform when the signal drives a single, isolated transition (e.g. from low to high). This step response is shown in Figure 81(a). The impulse response is the derivative of the step response, shown in

[122] Thus it is a very good tool for prediction data dependent jitter caused by the (passive) transmission path, and often also for driver effects as long as the source stays in its linear region. It is not a good approximation for clamped, limited or otherwise highly nonlinear systems.

[123] E.g. if the actual waveform transitions between 0 and V_{cc} then we can subtract $V_{cc}/2$ to get a signal usable for our theory.

Figure 81(b). One way to excite maximum positive response from the system is to drive positive signal (i.e. $+A$) at times when the impulse response is positive, and negative signal (i.e. $-A$) when the impulse response is negative. (If we drove instead the exact opposite pattern, we would get the maximum negative response). This worst-case pattern is shown in Figure 81(c). Those ranges in the impulse response correspond to the ranges in the step response where the latter has positive and negatives slope, respectively, separated by the local maxima and minima of the curve.

What is left to do now is to mark all the successive differences between these extrema ($\Delta_1, \Delta_2, \Delta_3, \ldots$, shown in Figure 81(a)). The absolute maximum output we can ever get from the system is the given by the sum of their *absolute* values, multiplied by A:

$$A_{out,max} = A \times \left(|\Delta_1| + |\Delta_2| + |\Delta_3| + \ldots \right). \tag{86}$$

The overshoot corresponding to this signal is then $\Delta A = A_{\max} - A$. We know that voltage errors (noise in the widest sense) and timing jitter are related through the rise time (or the slew rate) of the signal, so from this maximum overshoot we can determine the maximum possible peak-to-peak data dependent jitter value:

$$\Delta T_{pk-pk,\max} \approx 2 \times T_R \times \frac{\Delta A}{A}. \tag{87}$$

The factor 2 comes from the fact that we can produce an overshoot of ΔA as well as an undershoot of $-\Delta A$, and the timing errors, both approximately equal, add up to the peak-to-peak value.

We should note, however, that this is really an absolute worst case upper bound that we are unlikely to attain with any regular bit stream: The worst case signal will most likely not have equally spaced transitions (since the transitions are determined by the change in slope of the step response, which can be very irregular), so we can't produce it with our regularly-spaced edges of any well-behaved data signal. The method's value lies in the fact that this bound is easy to determine – all we need is a single step response measurement and a little bit of math, as opposed to going through a huge variety of different patterns in the hope of exciting every possible combination. And if it turns out that the theoretical upper bound is already small enough for the system to fulfill its requirements, then we have saved ourselves a huge amount of worry and work. If we need a better estimate for a particular data stream, we can always go back and perform a full-blown measurement as described previously.

Chapter #4

MEASUREMENT ACCURACY

10. SPECIALIZED MEASUREMENT TECHNIQUES

In a perfect world, all we would ever need for jitter and timing measurements would be a real-time sampling scope. Out of all instruments it is the most straightforward to use, it yields the most comprehensive information about the signal, it can do single-shot acquisition, the software on high-end scopes greatly simplifies jitter analysis, etc. But as we have seen its performance is limited in several respects – namely its acquisition speed (compared e.g. to BERT boxes), bandwidth and to some extent accuracy (compared to equivalent-time sampling scopes). So obviously the only good reason to use a different instrument is if we absolutely need to achieve either higher throughput, higher accuracy, or maximum bandwidth. In this chapter we will look at a few specialized measurement techniques on several instrument types that either yield superior accuracy compared to standard instrument performance (e.g. jitter measurements down to the femtosecond range on a scope with intrinsic jitter of picoseconds), or allows specific jitter measurements on instruments that originally were not intended to have this capability (e.g. periodic jitter measurement on a digital tester). The list is can in no way be comprehensive but is intended to showcase how with some engineering ingenuity one can push the performance and capabilities of an instrument far beyond its original specifications.

10.1 Equivalent-Time Sampling Scopes

10.1.1 Phase References

Recently there has been an interesting progress in the reduction of time base jitter on equivalent-time sampling scopes (since those types of scopes provide the highest bandwidth to look at the fastest signals, low instrument jitter is of utmost importance here). So-called precision time base reference modules[124] as an optional plug-in can improve the jitter performance to about 200 fs RMS, compared to a standard value that is around 700 fs or worse without the reference module.

How is this possible? The secret is that it needs a separate reference signal – assumed to be perfectly jitter-free – to which the scope will lock its time base. This signal has to be sinusoidal (or at least close to sinusoidal; a square wave will not work). Now this may sound like a vicious circle, but in fact quite often such a "perfect reference" is readily available in the shape of the system clock of the system under test – per definition this is the reference for all timings on the board. Or we can think of using a very low-phase-noise (i.e. low-jitter) sine-wave generator that can provide a more stable reference than the oscilloscope's internal timing system.

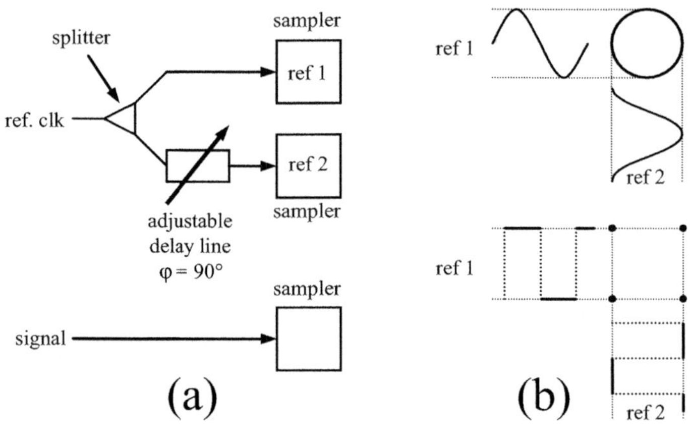

Figure 82: (a) Block diagram of a high-end equivalent-time sampling scope's phase reference. (b) Lissajous plot with two sine waves in quadrature (top) and two square waves in quadrature (bottom), respectively.

[124] The exact naming for those components is vendor dependent.

Leaving internal details aside, the jitter reduction process works as follows (shown in Figure 82(a)): From the single-ended reference input signal the scope generates two versions that are in "quadrature", i.e. they are delayed by 90 degrees (a quarter-period) relative to each other. If (just virtually) we use one signal for deflection along the x-axis, and the other for deflection along the y-axis, then the well-known Lissajous plot emerges (see Figure 82(b)). Each point on the plot corresponds to a well-defined phase within one cycle. This plot is what the scope acquires and stores for later use when the reference is first applied and the scope is told to characterize (lock on) the phase reference.

At each sampling instant the scope now samples not only the input signal to be measured, but also the two versions (shifted by 90 degrees to each other) of the reference clock. With the two phase reference samples it can go to the Lissajous plot and determine the exact phase within the cycle (and thus the exact time) that the sampling actually took place, with much better accuracy than the standard time base would be capable off.

However the true strobe jitter is still the same whether the phase reference is used or not – only now the scope knows precisely where a particular strobe event fell – so the scope needs to acquire several samples to get one close to the desired timing position (the other strobe results are thrown away). This reduces the usable sample rate by a factor of about 4 – the price to pay for reduced measurement jitter.

Looking again at Figure 82(b) it also becomes clear why a square wave does not work: There is no unique relationship between a pair of sampling values and the phase – the Lissajous plot degenerates to four discrete points at the corners. With a sine wave the plot is a circle. If our reference clock is not sinusoidal, then we need to filter it (with a rise time filter of appropriate size) to make it sinusoidal.[125] Second, the relative accuracy of the sampling gets better with better signal-to-noise ratio, which means it is advisable to use as large a reference signal as possible (up to the maximum input amplitude range of course).

In addition to reducing the random time base jitter, the phase reference method has the capability to improve the time base linearity as well.

10.1.2 Eliminating Scope Time Base Jitter

While the phase reference is a very useful addition to our tool set, there are many cases where the only scope available does not have this option; or

[125] Such a filter may already be integrated in the module. Otherwise a usable choice is a rise time of approximately $0.35/f_{CLK}$ so the desired clock frequency is only weakly attenuated (by 3 dB) but higher components strongly.

we may need an even lower jitter floor than 200 fs.[126] We will now look at a method that – based on experiments performed by the author – for certain situations enables jitter measurements down to below 40 fs RMS!

As we saw earlier, noise on a signal translates into jitter on a transition. But it works the other way around, too: If a signal has some jitter, than this will look like noise on a transition. Figure 83 illustrates this: Jitter makes the signal move left and right, and if always sampled at the same position in time, the acquired voltages will vary – the translation factor being the slope (in V/s) of the signal. Of course any "real" noise present on the signal will add to this (as RMS if both jitter and noise are random). But for the moment let's assume the signal is perfectly noise-free.

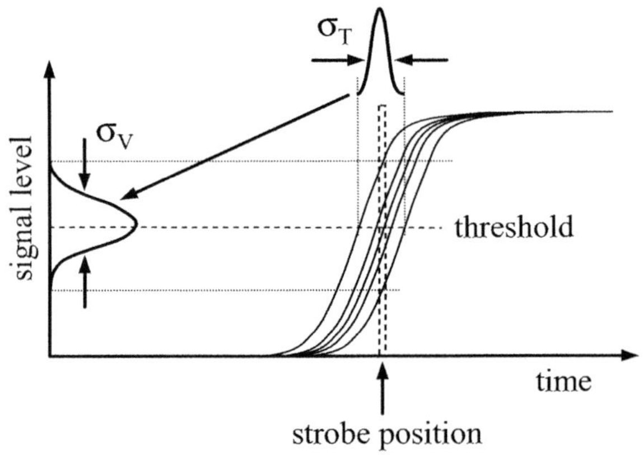

Figure 83: When a signal transition is repetitively sampled at the same timing location, jitter on the signal transition translates into sampling noise.

As our next step let's assume we have not just one, but two signals – the one to measure, and a reference signal. For simplicity let's say the two signals have same edge shape and same size, and that we can plot their difference:

$$M = signal1 - signal2 \tag{88}$$

What will we see? If the signals do not have any jitter *relative to each other*, and their transitions are lined up relative to each other in time, then the result will be a perfectly straight flat line. What if the two signals *do* have some jitter relative to each other? As Figure 84 shows, this will

[126] Many recent multi-Gb/s devices exhibit jitter at or below 200 fs RMS so even the best scopes are already being pushed to their limit there.

produce a peak in the difference signal whenever there is a transition because one signal will start rising (or falling) before the other does. The maximum peak height (at the center of the transition) is given by

$$H_{peak} = \frac{dV}{dt} \times \Delta T = S \times \Delta T \quad (89)$$

with S being the slope (in V/s) of each of the signals, and ΔT being the time delay between the transitions (caused by the jitter): As indicated in the beginning, timing jitter is converted into voltage noise.

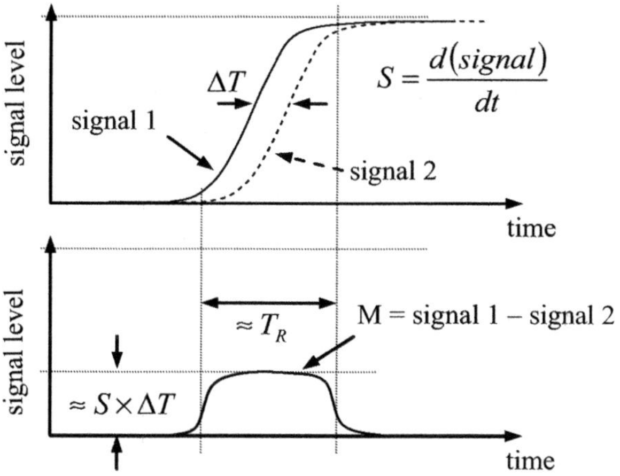

Figure 84: If two signals of otherwise same shape have timing differences between their transitions (e.g. because of jitter), the difference signal exhibits peaks during the transition regions.

The intriguing property is now that as long as we sample *somewhere* along the transition region where S is approximately constant, the result will be the same, no matter where exactly that sampling instant is. The width of that area is of the order of the signal rise time, as also indicated in Figure 84. In other words, even if our sampling point is somewhat off due to inherent scope time base jitter, this will have only negligible effect on the actual measurement result as long as the peak-to-peak jitter is much smaller than the signal rise time. *We have thus completely eliminated the effect of time base (sampling) jitter on our jitter measurement*! We can acquire a histogram of the noise on M at a single sampling instant and with the knowledge of S convert this directly into the jitter histogram, e.g. for the standard deviation:

$$\sigma_T = \frac{1}{S} \times \sigma_V, \tag{90}$$

where σ_T is the timing jitter's standard deviation and σ_V is the standard deviation of the histogram in the voltage (level) direction. The conversion formula for any other parameter (e.g. peak-to-peak spread) is identical.

So far we have made quite a few assumptions about the signals that may not hold in the real application:

1. The signals are of same size.
2. The signal rise times (or more precisely: the slew rates) are equal.
3. The transitions of the two signals are lined up in time with respect to each other.
4. The signals (and the sampler) don't have any noise.

We can resolve the first two restrictions easily if we realize that what really counts for the method to work – the noise peak being independent of the sampling instant – is only that the *slew rates* are equal. Thus, if the two signals have differing slew rates[127], S_1 and S_2, we can always have the scope scale the reference waveform S_2 so its slew rate matches the slew rate of signal S_1: In other words we must plot

$$M = signal1 - \frac{S_1}{S_2} \times signal2 \tag{91}$$

and then perform the acquisition and processing as described before. The conversion equation remains virtually the same:

$$\sigma_T = \frac{1}{S_1} \times \sigma_V \tag{92}$$

There is no such easy way out for the third requirement – we really have to line up the transitions on the two signals on top of each other to within a fraction of the signal rise time. Many higher-end scopes offer deskew features where the user can move one trace with respect to the other. *But beware*: If we want to measure random or other uncorrelated jitter (which will be different from one sampling instant to the next!), we *must* acquire both channels simultaneously – we cannot have the scope interleaving the

[127] Because the signals' rise times and/or swing amplitudes are different.

two acquisitions. Unfortunately many scopes don't do this at all or cannot do this when deskew is turned on. So what are our options?

First we could use a true differential probe and feed the two signals into its two inputs. The scope then needs only a single channel to sample the difference signal. The downside is that first – being an active element – the probe will inevitably add some additional noise and random jitter (we could circumvent this by using a passive Balun instead), but much more important, it does not allow to scale the signals if their slew rates are different from each other.

Second we can delay the signals by using high-quality trombone line stretchers to adjust the relative edge timing. Those elements provide down to sub-ps resolution – largely sufficient because we only need to adjust the edges so they are not more than a fraction of a rise time apart, which even for very fast signals means several ps.

"A fraction of the rise time" corresponds to "a bandwidth several times higher than the one given by the rise time", and it is highly probably that the signal rise time is somewhere close to the minimum rise time (maximum bandwidth) that our system under test (e.g. receiver) can handle anyway – noise at higher frequencies will not go through anyway but get smeared out, while the (higher-performance) scope is still able to show it. So aligning to a fraction of the rise time means the jitter acquired on the two signals is really from the "same" instant for all practical purposes.

Finally, for maximum accuracy and minimum jitter floor, we may want to take out the effect of the sampler noise. This noise adds up as root-mean-square to the noise produced by the jitter (assuming both are Gaussian, random events). We can determine the sampler's noise by disconnecting the signals, terminating both inputs with 50 Ω terminations, and measuring the RMS noise $\sigma_{V,sampler}$ on the difference signal M. The net RMS timing jitter σ_T is then approximately

$$\sigma_T = \frac{1}{S_1} \times \sqrt{\sigma^2_{V,\,total} - \sigma^2_{V,\,sampler}}\,. \tag{93}$$

We haven't yet talked about where to get the reference signal S_2 from. One possibility is of course that we have some reference like the system clock or a very low-jitter RF generator. But often such a reference is not readily available, so we need to look elsewhere:

Maybe all we really want to know is how much a particular element (e.g. a driver or an amplifier) in our signal chain is *adding* – i.e. we want to measure its *jitter insertion*. This is per definition a relative measurement where we want to compare the jitter of the signal coming out of this element to the jitter of the signal going into this element. So the former is our signal S_1 and the latter is the reference signal S_2. We will need to delay S_2 by the same amount as S_1 is delayed by the element under test so the edges line up and we really compare equivalent edges to each other. The whole setup is shown in Figure 85(a).

Finally we can use our method to look at edge-to-edge jitter (jitter of one edge to the next, or a few edges away – in other words, period jitter or N-period jitter), with a setup schematically shown in Figure 85(b). Our reference here is the leading edge, and the signal to compare it to is some later edge. To do this, we need to split up the signal and then delay one branch by an integer number N (at least 1) of cycles. The nice feature is that we can do this on virtually any signal, even if we have no other reference;[128] of course it can only show us short-term jitter, up to the maximum delay which is limited by the longest cable we can send the signal through and not attenuate it too much.

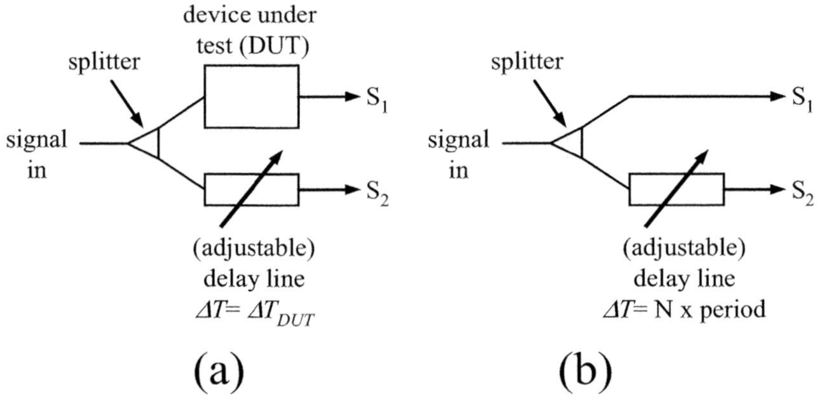

Figure 85: Eliminating the effect of scope time base jitter by using a reference signal: (a) setup for jitter insertion measurements, (b) setup for N-period jitter measurements.

10.1.3 Time Interval Measurements

10.1.3.1 Time Interval Errors

When we talk about timing measurements on an oscilloscope, what we usually refer to is the timing relative to some reference (e.g. the clock edge,

[128] As long as the delay element does not add too much distortion, rise time degradation, or attenuation, the slew rates S_1 and S_2 will automatically be almost identical.

the trigger, or some other data signal edge), but not any "absolute" point in time. Thus, what we are really interested in is the time delay (or time interval) between two events. The distance from any trigger is much less important (unless this trigger is our reference, but even in this case we could split off part of the trigger signal and display it as a second signal on the screen.

What is therefore of vital importance for any timing measurement is the accuracy of the oscilloscope's time base, because this enters directly into the time interval measurement accuracy. Errors of the scope can be of two kinds: random (different for every repeat) and systematic (the same for every repeat under the same conditions). Random errors can usually be averaged out if we have a stable, repetitive signal, but how about systematic errors?

Any oscilloscope (or timing measurement device in general) has to generate its so called "time base" which it uses to determine when exactly it must strobe (measure) the incoming signal. It usually does this by counting periods of some master clock, plus interpolation or delay within the clock cycles to obtain fine timing resolution. But even if the master clock is perfectly stable, interpolation or delay circuits – as any analog element – always have some amount of nonlinearity, and even calibration (often called "compensation") will leave residual errors.

All this is of little concern for real-time sampling scopes, where the acquisition is running continuously, asynchronously to the signal. As long as the internal scope period is not an exact multiple of the signal period (an unlikely case, especially given that there will be rather random breaks in the acquisition for data transfer and data processing), chances are the timing events will fall on a different place on the interpolator every time. Nonlinearities will thus show up as random scope time base jitter and we can average them out if we are only interested e.g. in data dependent effects. It is of course vitally important for properties like random or uncorrelated periodic jitter where we cannot improve the accuracy by averaging.

10.1.3.2 How to Spot Time Interval Errors

The situation is different for typical equivalent-time sampling scopes, usually the tool of choice whenever highest bandwidth and highest accuracy is required. Here it is the trigger that starts the delay timing measurement that leads to the acquisition process, so at the same delay after the trigger we are also at the same place in the interpolation curve. In other words, the timing error for a given signal is determined by its distance to the trigger event.

There is an easy test for the approximate size of such errors. Assume we have two signal edges, i.e. a single pulse, and we set the scope up so it measures the delay between them (for a pulse this would be the pulse width).

As long as we don't do anything to the signal itself, this pulse width is obviously constant. If the trigger occurs at time t_T, the first edges occurs at time t_1, the second at t_2, and the time base errors at t_1 and t_2 are $\varepsilon(t_1 - t_T)$- and $\varepsilon(t_2 - t_T)$, respectively, then the displayed delay (time interval) Δ and its corresponding error ε_Δ is

$$\Delta = (t_2 + \varepsilon(t_2 - t_T)) - (t_1 + \varepsilon(t_1 - t_T)) = (t_2 - t_1) + (\varepsilon(t_2 - t_T) - \varepsilon(t_1 - t_T)), \tag{94}$$

and

$$\varepsilon_\Delta = \varepsilon(t_2 - t_T) - \varepsilon(t_1 - t_T). \tag{95}$$

If we can move the *trigger* timing t_T in small increments, then we are guaranteed to leave the real signal itself unchanged, and any change in the *measured* delay (averaged to get rid of noise and random jitter) can only be due to the oscilloscopes time base nonlinearity that is different for each position. Of course if we pick a pulse width that corresponds to an integer multiple of the scope's internal period, we won't observe any variation because the systematic (non-random) interpolation errors are largely periodic with the cycle period. It is therefore a good idea to try this with a few different periods that are not in a simple ratio to each other. The latest high-end equivalent-time sampling scopes still have linearity errors of a few ps, so this can be of concern for very accurate measurements of high-speed signals (at 10 Gb/s, i.e. with a bit period of 100 ps, an error of 10 ps is 10% of the unit interval!).

10.1.3.3 Sub-Picosecond Time Interval Accuracy

Luckily, there is a special averaging method that can reduce this error and allows for sub-ps time interval measurements even on present-day oscilloscopes:

It is reasonable to assume that that master clock in the oscilloscope is optimized for high stability, so the main contribution to any time base errors will rather come from the analog interpolator (or delay) element and not from the clock. In other words we assume that the timing error from the beginning of one scope clock cycle to the next is zero, with deviations only within the cycle. If we move the trigger timing over exactly one full cycle in small steps, each time acquiring the signal edge position(s), the errors will assume a variety of different values, but overall they will average to something close to zero. The smaller the steps, the better the average (remember the inverse square-root behavior with increased sample size!). If we want to be on the safe side and also account for variations in the clock period, then we can do the same over several clock periods if we can afford the increased acquisition time. The only thing that is important is that we move the strobe over an *integer* number of clock periods, and not e.g. 2.3 periods wide.[129]

For a scope with a 10 ns master clock period the measurement algorithm could be as follows:

1. Set up signal averaging (use a sufficient number of averages to reduce random signal and time base jitter to less than the required measurement accuracy)
2. Set up trigger and signal.
3. Measure the signal timing(s) of interest.
4. Move the trigger timing by 100 ps.

Repeat 3. and 4. until the trigger has been moved by $10\,\text{ns} - 100\,\text{ps} = 9.9\,\text{ns}$, i.e. 100 repeats in total.

A rough guess for the expected accuracy enhancement factor is then $\sqrt{100} \approx 10$ assuming the errors are more or less randomly distributed, so with the specified accuracy of this type of oscilloscope of (5 – 10 ps typical) we can expect an absolute time interval accuracy of better than 1 ps!

In fact, the number of averages set up in step 1 can be smaller since we average all the curves (100 in our example) together, which provides additional averaging, so we could reduce them by $\sqrt{100}$ and still minimize random jitter sufficiently.

10.2 Digital Testers and Bit Error Rate Testers

10.2.1 Edge Searches and Waveform Scans

As we have already seen in section 2.6, a comparator in a digital tester can only tell us if the incoming signal was high or low in a given cycle, but it neither tells us when exactly the transition happens nor what the exact level

[129] Now the internal clock period is not something that is usually contained in a scope's data sheet, but fortunately the number of high-end, high-accuracy scope models where this extreme accuracy makes sense is limited anyway. For Tektronix' equivalent-time sampling scopes (11800 series, TDS/CSA8000 series) the period is 2.7 ns, while for the latest Agilent models (86000 series) it is 4 ns.

at the strobe position is. On the other hand it is usually easily possible to re-run the pattern (and thus the measurement).

To acquire the exact edge timing, we could thus repeat the run over and over again, each time moving the strobe timing by a little bit (high-speed testers allow resolutions of 10 or even 1 ps), as shown in Figure 86(a). Also displayed are the selected threshold level and the result of each strobe. When – as assumed in Figure 86 – the signal has a rising edge in this timing range and the strobe is moved to successively later times, the compare results will first all be "low" (because the edge has not yet transitioned when the comparator strobes), and when the strobe position crosses the edge position the subsequent results will all say "high". In this way we can determine the exact crossing point with the minimum resolution of the tester's timing system.

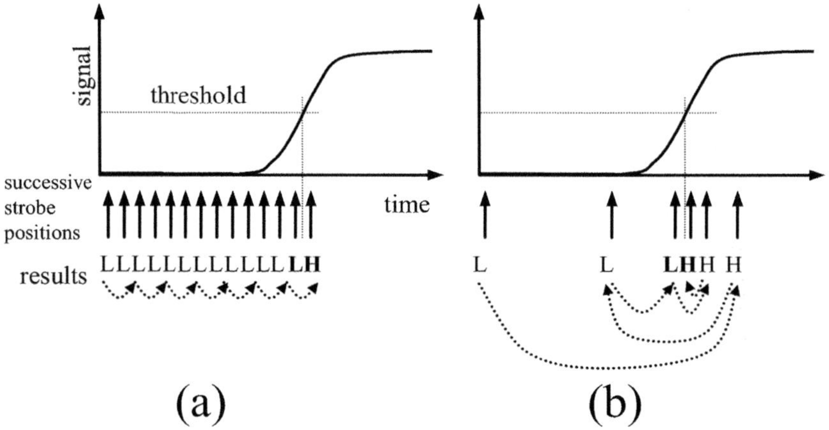

Figure 86: (a) Linear edge search with a comparator set to the threshold voltage. (b) A binary edge search (successive approximation) finds the edge much faster than a linear search.

If instead we want to know the voltage at a given timing instant, we would perform a very similar repetition, but this time leaving the strobe position unchanged but move the threshold level – this way we basically digitize the waveform at the given instant with high resolution. The searches will take proportionally longer and longer the finer the desired resolution gets, so an obvious improvement is to use binary searches (successive approximation) instead of the linear scans – the number N of repeats to find the edge (or the level) with a given resolution in a certain search range is then only approximately

$$N \approx \log_2 \frac{resolution}{range}. \tag{96}$$

Another way to see this – to enhance the analogy with oscilloscopes – is that to get N bits of effective resolution we need N steps in the search. The algorithm for binary searches is graphically illustrated in Figure 86(b) for a search in the timing direction (a voltage search would work identical) :

1. Determine result R_l at lower end of the timing range (T_l)
2. Determine result R_u at upper end of the timing range (T_u)
3. Determine the result R_m at the midpoint ($T_m = (T_u + T_l)/2$)
4. If $R_m = R_u$, then set $T_u = T_m$, otherwise set $T_u = T_m$
5. Repeat 3. and 4. until the range (T_u-T_l) is smaller than the desired resolution.

For example, for a signal that varies from 0 to 1 V and that we need to digitize with 1 mV resolution, the number of repeats is approximately 10. This is a huge improvement compared to the linear search that may take 1000 steps of 1 mV each if the signal happens to be at the end of the search range.

If we need to digitize a full section of the waveform (i.e. acquire data equivalent to an oscilloscope), then we can simply step the timing through the desired range, at each position digitizing the voltage with a binary search. While not exceedingly fast with regard to acquisition rate, this has made our humble digital tester into a full-blown equivalent-time sampling oscilloscope! (of course except for the fact that the scope needs only a single shot to acquire the voltage at a given instant).

10.2.2 Signal Rise Times

To get the rise time of the signal, in principle all we have to do is two edge timing measurements (binary searches) at the two levels of interest (e.g. 10% and 90%); it may also be necessary to look at static high and static low levels coming from the signal source to establish the low and high baselines (i.e. 0% and 100%). To average out random timing jitter we need to repeat the timing measurements several times.

The biggest obstacle here is that as mentioned in the previous chapter digital comparators are normally not trimmed to very high signal integrity. After all, their intended application if to do a rather coarse high-low decision in the center of a bit interval, i.e. far away from the transition at a time when the signal should have more or less settled, so the analog performance is more likely than not "just good enough for the task". Limited analog input bandwidth, distortion of the incoming waveform and slow settling behavior (long, complex, lossy path between device and comparator) should keep the expectations in the accuracy of high-bandwidth measurements (as is any rise time measurement) realistically low.

10.2.3 Random and Data Dependent Jitter

With the toolset from the previous chapter we can in principle already do a rudimentary timing and jitter analysis on our signal.

If we repeat the edge search (i.e. binary search in the timing direction) on the same edge in the pattern many times, we will find that – as long as our resolution is fine enough – due to random jitter the results will vary from one repeat to the next. We can then apply statistical analysis to the results' distribution to get the average edge position (the mean of the distribution) as well as the RMS or peak-to-peak jitter.

Going from one edge to the next, we can collect timing information on each edge and build up the data necessary to determine data dependent effects or duty cycle distortion. If the single-edge distributions are close to Gaussian bell curves, then we can attribute their width to random jitter.

But the main downside of this approach is that it is very time consuming because of the huge number of patterns repeats necessary – each measurement takes several search iterations, each edge needs to be measured several times (at least about 100 times just to get reasonably reliable RMS jitter numbers), and there may be many edges in the pattern. It may be feasible in a characterization application where runtime is of secondary importance, but certainly not under production test conditions.

Also the timing accuracy and linearity (summed up in its edge placement accuracy or "EPA") of a production tester can never match a good high-end oscilloscope or similar device, so the absolute timing values may be off quite a bit. As a rule of thumb (even though it would be better to consult the tester's data sheet regarding timing accuracy), linearity deviations will probably be at least a few times the minimum timing resolution, and random jitter may be of the same order of magnitude. The first will affect average edge timing measurements, while the latter impacts directly the random jitter numbers.

One more weakness is that it can be difficult to distinguish other, deterministic effects in all cases. E.g. periodic jitter whose period is not related to the bit period will modify the single-edge distribution to a non-Gaussian shape, but we can't get information about its frequency – and thus, even worse, we can't even tell if it is really periodic jitter, and not rather some other kind of uncorrelated deterministic jitter. To even detect deviations from a Gaussian distribution reliably, we need a large number of samples. The previously mentioned 100 repeats are good enough if all we want is mean position and standard deviation, but is far too small to accurately represent the full two-dimensional distribution: e.g. if the distribution is just 20 resolution steps wide, and we have 100 samples, then each bin of the PDF has only 5 hits on average. Statistical shot noise (in the present example equal to about $\sqrt{5} \approx 2.2$ (and also aliasing if the resolution is not much finer than the distribution width) will cause so much statistical noise that there is no way of telling with any confidence if the distribution is really non-Gaussian or if the deviations are merely statistical noise. To overcome this, we need maybe about 100 samples *per bin* of the PDF, so for a distribution 20 bins wide this increases the necessary amount of data to 2000 repeats – proportionally increasing the acquisition time.

10.2.4 Coherent Undersampling

One way overcome the acquisition speed and accuracy limitation of a digital production tester – without taking refuge to external instruments – is a method often called *Probability Digitizing* or *Coherent Undersampling*. Its principle of operation is illustrated in Figure 87.

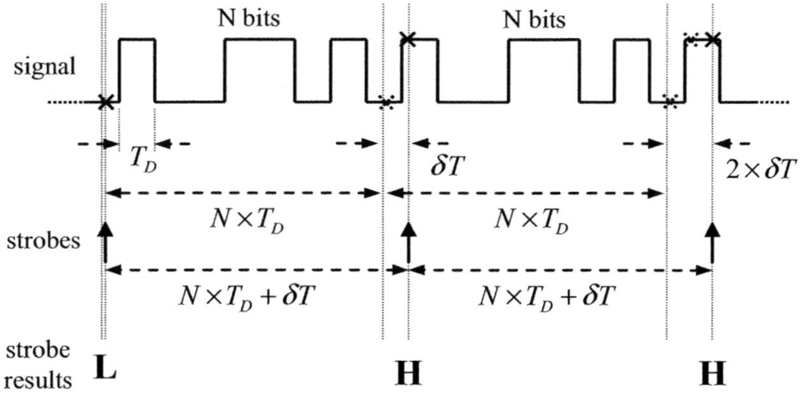

Figure 87: Coherent undersampling using a sample clock period very close to – but not identical with – the bit period.

A fundamental requirement for its implementation is that the tester must be able to run in at least two different "clock domains", i.e. there are two groups of pins that run at different cycle times. Since many modern devices require such capability anyway (e.g. a processor running at some core speed but having a memory bus at a different speed), this is a more and more common feature on production testers today.

With that capability we can then run all the input data pins and the clock to the device at some bit period T_D, while we run the compare pins (that receive and strobe signals coming from the tester) at a slightly different speed

$$T_C = T_D + \delta T, \quad \delta T << T_D . \tag{97}$$

We let the pattern (consisting of N data bits) repeat over and over again without interruption, and for the sake of simplicity let's assume for the moment that we place only a single strobe in the pattern. At the first pattern repeat, the strobe occurs at some timing position T_1 within the pattern. At the next repeat, it occurs slightly later, namely at

$$T_2 = T_1 + N \times \delta T , \tag{98}$$

and the next repeat will be even further into the pattern by the same amount, and so on. This is very similar to the way an equivalent-time sampling scope works, except that the comparator only distinguishes between high and low, but does not tell us the exact voltage at each point. If we make the two periods *very* close to each other (it often is possible to have them differ by only a few fs), the resolution – or equivalent sample interval – becomes extremely fine. We can then consider several subsequent samples to have occurred at virtually the same time (e.g. if our resolution is 10 fs and the time scale of interest is 1 ps, we would lump every 100 results together), which is equivalent of repeating the same run many times in the previous section – so statistics over those groups of samples gives us random jitter information. In reality we would probably also place a compare strobe into each cycle, and as we already mentioned the tester can capture the pass-fail results in real time and store it in its fail vector memory, so the acquisition rate is immensely fast – e.g. 1 GSample/s if we run at 1 Gb/s (figuring out which strobe corresponds to which equivalent timing position in the data pattern requires some number crunching but can be done automatically in software).

Using this method we avoid moving the *programmed* strobe positions within the cycle, which in turn avoids edge placement accuracy issues (limited resolution and linearity) mentioned before; in most test systems the cycle period is something much more stable (otherwise the fine period resolution would not make sense to begin with), so we end up with a hugely refined timing accuracy as compared to the brute-force approach of the preceding section.

The second name of the method (probability digitizing) comes from the fact that transition probability (more exact, the CDF of the waveform) gets translated into a stream of digital bits: When the strobe position approaches (in steps of $N \times \delta T$) an edge in the waveform, it moves into the range where the edge – depending on random jitter (or some other uncorrelated effect like periodic jitter) – may or may not have transitioned yet. The further the strobe moves, the more probably it becomes that it already transitioned, so the percentage of resulting "ones" (of it is a rising edge) will increase, in proportion to the CDF of the edge. Of course subsequent strobes are in reality one pattern repeat apart, but their *equivalent* timing position is very close to each other.

The information that we can extract is exactly the same as in the previous section (albeit with much better accuracy), so e.g. there is no way to determine the frequency of periodic jitter. Also the considerations regarding the necessary amount of data per edge – to accurately reconstruct the statistical edge timing distribution – stay the same.

10.2.5 Spectral Error Density Analysis

The good news is that we can take the basic idea of probability digitizing and modify it a bit to employ to identify additional jitter contributors – namely periodic as well as uncorrelated deterministic jitter.

For simplicity, let's assume our data stream is a simple clock-like pattern (which is good enough since we can already determine data dependent effects with standard probability digitizing). This time we do not move the strobe through the pattern, therefore there is not even a need for a second clock domain, and the method is applicable to virtually any digital tester without restriction. Instead, we place the strobe at the average edge position (the center of gravity of the edge distribution).[130]

First, let's look at what happens if the data stream (and the strobe as well) has no random jitter present whatsoever, but still some periodic jitter: If the latter moves the edge to the left (earlier) at some interval of time, then all strobes during that interval will yield "pass" because the edge has already transitioned before the strobe. On the other hand, if in some interval the periodic jitter has moved the edge to the right (later), all strobes will yield "fail". This is illustrated in Figure 88(a). As the end result, the captured bit

[130] Note that in a "normal" transmission application the strobe gets typically placed in the *center* of the bit period (not at its limit as done here), i.e. as far a way from preceding and following transition as possible, in order to minimize the BER.

stream will show alternating blocks of "pass" and "fail", respectively, with a rate of change equal to the frequency of the periodic jitter. What we don't know yet is the size (peak-to-peak deviation) of the periodic jitter.

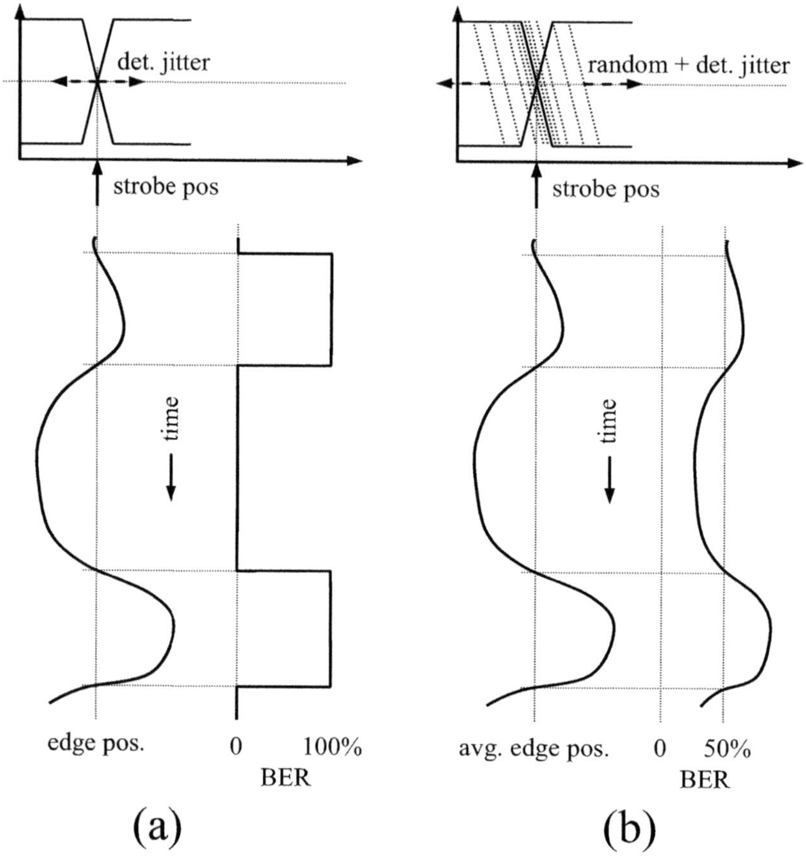

Figure 88: Spectral error density analysis on a digital tester or a BERT box: (a) ideal deterministic jitter without random components, (b) random jitter dominating the deterministic jitter.

Any real signal will have some random jitter in addition. In this case, if the periodic jitter moves the (average!) signal a bit earlier, the fail count will not immediately drop to zero, but just get somewhat smaller, since even here the random jitter at some specific strobe may make the edge late enough to fail. But the further the average position is moved earlier, the less fails we will get. The resulting time trend of the BER is indicated in Figure 88(b). If we know the random jitter distribution (histogram), we can therefore map the observed fail count (in other words, the bit error rate) to the size of the displacement (due to periodic jitter). As long as the latter is small against the random jitter's distribution width – approximately $\pm\sigma_{random}$ – the random jitter PDF is roughly constant and the mapping (displacement vs. change in fail count) is approximately linear since it is virtually identical to the edge's CDF[131] (see Figure 50).

We can then take the captured bit stream and do a spectral analysis (Fourier transformation) on it, and this gives us all the frequency components of the periodic jitter: They will stand out as peaks from the spectral plot, with the peak area corresponding to their peak-to-peak timing deviation. This even works when there is more than one frequency component present.

The only limit is the by know well-known Nyquist theorem – the sample rate must be at least twice as high than the highest frequency component of the periodic jitter, higher frequencies are aliased (folded back) back into this range and give erroneous results. But given that a tester can sample once per bit period, this means no great restriction. – E.g. when running at 1 Gb/s we can correctly detect jitter components up to 500 MHz.

10.3 Spectrum Analyzers

As described in the previous chapter, unlike all other instruments that we looked at, spectrum analyzers work in the frequency domain, while we are interested in the time domain. Thus they may seem unlikely candidates if we need to make timing (i.e. not frequency) measurements on digital signals. However, as we will see below, a spectrum analyzer can be a very powerful and flexible tool for two types of jitter, especially on periodic (clock) signals: for periodic jitter and for random jitter.

Let's consider a periodic signal of some frequency f_0. If the signal is purely sinusoidal, then its spectrum (as displayed on the spectrum analyzer) is a single narrow peak centered at frequency f_0. (In theory, the peak would be infinitely narrow and infinitely high, with a total area proportional to the signal amplitude. But since every real-life spectrum analyzer only has a finite resolution bandwidth, the displayed peak is at least one such bandwidth wide. However, the total peak area is not affected.)

If the signal is not sinusoidal, but e.g. – like it is often the case for digital clock signals – a square wave, then the spectrum changes to a series of peaks corresponding to its Fourier components, at frequencies f_0, $2f_0$, $3f_0$, and so on, as shown in Figure 3 for the example of a square wave. We then call the first peak the fundamental (or sometimes the "carrier") and all other

[131] If the random jitter is too small, we could think of artificially inserting additional random jitter through the clock signal going into the device.

"harmonics". In most cases the fundamental is the strongest component. Since many available spectrum analyzers can be set to very fine resolution, we can thus use those instruments to perform a very precise period measurement of our signal.

10.3.1 Periodic Jitter Measurements

Things become interesting when the period is not perfectly stable. In RF applications one would talk about "modulation", but in the time domain of course we call this "jitter". Nevertheless we will profit from RF modulation theory to determine the effects on the frequency spectrum.

For simplicity, we assume that the clock signal $a(t)$ looks sinusoidal. We can then express modulation (aka jitter) as a time-varying phase factor $\Phi(t)$:

$$a(t) = a_0 \times \sin\left(2\pi f_0 t + \Phi(t)\right) \tag{99}$$

The simplest case is when the modulation is sinusoidal as well, and the peak phase deviation is small compared to the signal period, i.e.

$$\begin{aligned} \Phi(t) &= \Phi_{\max} \times \sin(2\pi f_m t) \\ \Phi_{\max} &<< 1 \end{aligned} \quad . \tag{100}$$

$\Phi_{\max}$ is also called the "modulation depth". In this case the spectrum shows two sidebands to the left and to the right of the central peak f_0, at a distance equal to the modulation frequency f_m. This is displayed in Figure 89(a).

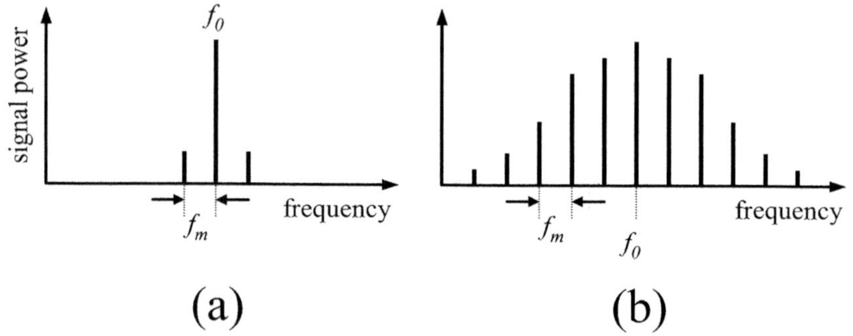

Figure 89: Spectrum of a signal with sinusoidal (single-tone) phase (or frequency) modulation: (a) small modulation depth, (b) large modulation depth.

The height of the sidebands is approximately proportional to the modulation depth. Going in the opposite direction, given a measured spectrum one can determine both the modulation frequency (from the spacing of the sidebands) and the modulation depth (from the height of the sidebands). Putting both together one can then calculate the timing jitter. As long as the phase modulation is caused by unwanted jitter on the signal (e.g. coupling in from a power supply), the condition of small modulation depth is likely to be fulfilled, so above relations are valid. (Larger modulation depth would most likely mean that the signal is not usable as a clock to begin with).

If the modulation depth is larger (e.g. because we artificially injected a large phase modulation on the signal), then we get a series of sidebands, all spaced by f_m from each other, as shown in Figure 89(b). The relative amplitudes of carrier and all harmonics is then given by the Bessel functions J_n (where n = 0 means the carrier and n = 1, 2, 3,... denote the sidebands). In the general case there is no way of giving mathematically exact values, instead those functions have to be calculated numerically.

A similar effect arises when the modulation is not sinusoidal. In this case we can decompose the modulation into a discrete Fourier series of sinusoidal modulation components (f_m, $2f_m$, $3f_m$, ...), and each of them will cause its own series of sidebands. Since the frequencies higher-order sideband of those components coincide (e.g. the second-order sideband of f_m lies at the same frequency as the first-order sideband of $2f_m$), determination of timing jitter from such complicated spectra soon becomes unfeasible unless the modulation depth is small (so each component produces only first-order sidebands.

If the clock signal is not sinusoidal, then each of its Fourier components (f_0, $2f_0$, $3f_0$, and so on) is modulated on its own. As long as the clock frequency f_0 is large against the modulation frequency f_m, those spectra remain well separated and can be processed independently from each other.

10.3.2 Random Jitter Measurements

As a second step, let's assume that we have some random timing jitter on the clock signal. Just like for non-sinusoidal modulation we can decompose the jitter into its frequency components, but now we get a continuous Fourier spectrum instead of discrete lines. Another way to visualize this is to think of the continuous spectrum as an infinite number of infinitely weak discrete components. We already know that for weak modulation the signal spectrum only shows two sidebands, spaced by f_m apart from the carrier, and "infinitely weak" certainly qualifies as "weak". Thus the random jitter causes the carrier peak to broaden – the stronger the jitter, the broader the peak becomes. This is shown in Figure 90(a) Again, if the clock signal is not sinusoidal, each of its Fourier components will experience the same broadening, as indicated in Figure 90(b). The center of the peak corresponds to the (average) period of the clock signal. The lobe areas close to the peak correspond to very low frequency jitter, while the areas far away a due to high-frequency jitter components. This means in the analysis we can selectively include only certain ranges of the jitter frequency in the calculation of the total jitter. E.g. we may not care about very low-frequency components that our system PLL will follow (compensate) anyway. Also, the spectrum analyzer has only finite resolution (e.g. 1 kHz), so components slower than that will only show up in the central peak. The total jitter is the ratio between the power in the side lobes and the power in the central peak.

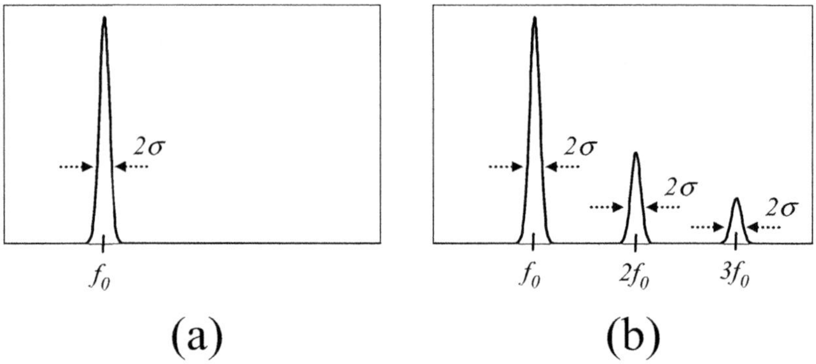

Figure 90: Broadening of a spectral peak due to random jitter (or, in the frequency domain, called phase noise): (a) sinusoidal carrier signal, (b) non-sinusoidal, periodic signal.

In the actual calculation one has to make several corrections to account for the way the spectrum analyzer works when measuring noise-type signals: First, since the analyzer display is logarithmic, it will over-emphasize small signals while attenuating large peaks in the signal (in other words, the average of the logarithm is not the same as the logarithm of the average over a signal). This causes an under-response of 2.51 dB which we have to *add* to the measured values.

Second, taking the integral over the curve implicitly assumes that each displayed frequency data point is independent of its neighbors. But this would require a "rectangular" bandpass filter whose response is flat over its resolution and falls off infinitely fast outside its pass range, while real spectrum analyzer filters fall off much slower. This causes "crosstalk" between different frequency bins. The correction factor, which we have to *subtract* from the measured values, depends on the exact shape of the filter response curve. Most analog spectrum analyzers have 4-pole filters with a correction factor of 0.52 dB, some have 5-pole filters (0.46 dB), whereas the factor for most FFT-based instruments is 0.24 dB.

Finally we have to normalize the data to a (virtual) filter bandwidth of 1 Hz by *subtracting* $10 \times \log(BW)$ where BW is the instrument's filter bandwidth.

The RMS jitter is then calculated as

$$\sigma_{jitter} = \frac{1}{f_0} \times \frac{\sqrt{2 \times \int_{f1}^{f2} S(f)df}}{2\pi}. \tag{101}$$

S(f) is the signal's power spectrum, f_1 is the starting frequency (frequency closest to f_0), f_2 is the farthest frequency of interest. This integration is also indicated in Figure 91(a), where the frequency is referenced to the mean frequency (carrier frequency) of the signal – a better visualization of the jitter effects than a plot over the absolute frequency. The factor of 2 under the square root comes from the fact that integration is done only on one side lobe of the peak, whereas in reality we have equal noise power in both side lobes.

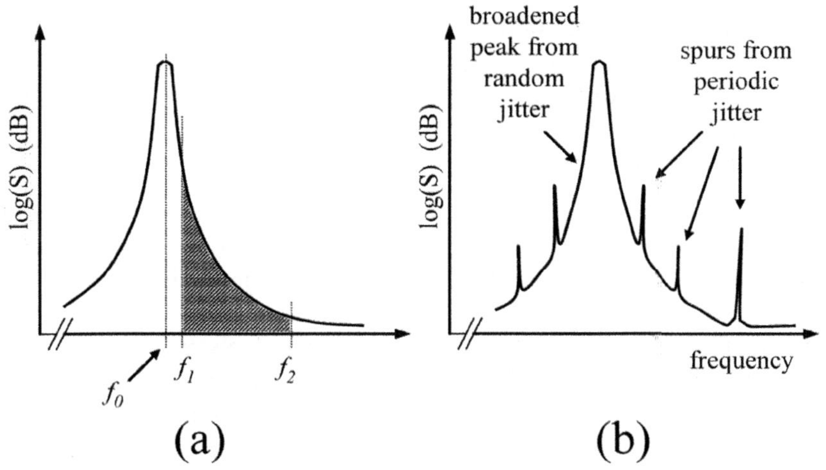

Figure 91: (a) Calculation of random jitter by integrating the power in the spectral peak's side lobes. (b) Periodic jitter components cause spurs (peaks) in the side lobes that have to be treated separately.

This formula is based on the so-called *Parseval Theorem*, a mathematical relationship between spectral representation and time representation, and it *assumes* a Gaussian jitter distribution (in other words, without further knowledge about the signal this may or may not be the case). Unfortunately the signal spectrum yields only information about the amplitude and the frequency of the jitter components, but none about their phase. This makes it impossible to reconstruct the actual jitter trend and thus the peak-to-peak jitter, unless – as the formula does – one makes certain assumptions (e.g. that the jitter is Gaussian).

If we have a mixture of random jitter and periodic jitter, then the periodic jitter will show up as discrete peaks in the spectrum (see before), as illustrated in Figure 91(b). A good analysis program is able to separate those peaks from the random spectrum and then process them separately from the random components.

The "phase noise measurement" using a spectrum analyzer has certain significant advantages over direct (time-domain) measurements:

- Good, high-end spectrum analyzers have unmatched signal-to-noise ratios, so even minuscule amounts of random jitter can be measured accurately (important for modern low-noise clock sources). That makes them the tool of choice when the jitter to be measured is far below the internal timing jitter of even high-end oscilloscopes (depending on setup, somewhere between maybe 200 fs and 1 ps RMS jitter).
- High frequency resolution allows us to resolve the spectrum close to the peak. This means very low-frequency jitter can be captured and analyzed – this would need excessive amounts of storage memory on a sampling scope in order to capture a sufficiently long data stream (remember that time and frequency are inversely related).
- The ability to mask out certain ranges of jitter frequencies by integrating only over the frequency range of interest. This way we can simulate the behavior and reaction of a PLL in the circuit driven by the signal.

Above jitter measurement method works extremely well for sinusoidal clock signals, so people in the RF test world are fond to use it, e.g. to characterize highly stable crystal oscillators. But for "digital people" dealing with data signals it has some serious shortcomings: First of all, digital signals (even clocks) rarely look sinusoidal but rather resemble square waves. This means they have strong harmonic content at integer multiples of the data rate (or clock frequency), which limits how far out we can do the spectral integration (because adjacent harmonics interfere). This is still not too bad because it allows at least integration up to about half the frequency.

Where the method really breaks down is at the analysis of data patterns (as opposed to clocks). In many cases data patterns over longer periods tend to look random, and as we have seen in Figure 5(b) in this case there are no well defined peaks in the spectrum – they are completely washed out.

10.3.3 Maximum-Accuracy Phase Jitter Measurements

Let's recall the section about random and periodic jitter measurement with a spectrum analyzer. While theoretically a perfectly applicable method, its practical limitation lies both in the achievable frequency stability of the analyzer's bandpass filter as well as in the noise floor of its amplification and filter system – remember that we *integrate* over a certain frequency band, so even small instrument jitter (when dealing with spectrum analyzers we speak of noise floor, but it works out to the same) when integrated over a wide enough range accumulates to a significant measurement error. This affects especially high-frequency jitter that spans a wide range away from the carrier frequency. At the same time the minimum frequency offsets from the carrier frequency for low jitter frequencies are very small compared to the carrier frequency (remember that the sideband separation for phase modulation is equal to the modulation frequency).

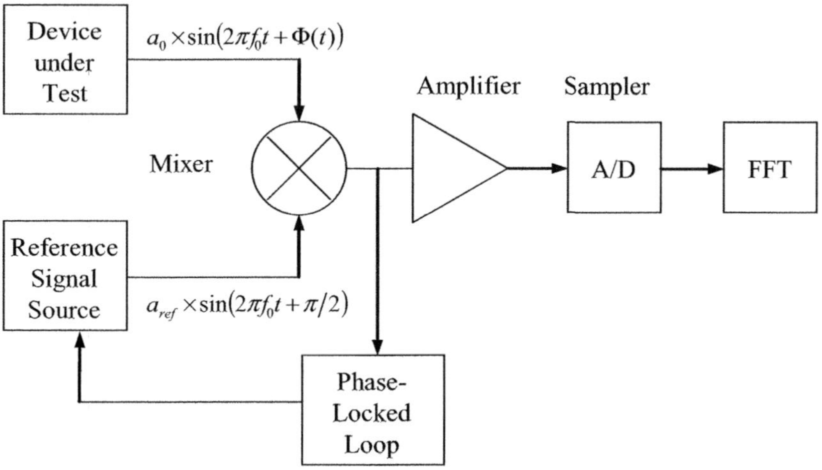

Figure 92: Simplified block diagram of a dedicated phase noise measurement with ultra-low noise floor.

For all those reasons there exist dedicated phase jitter measurement units that seek to improve on the basic spectrum analyzer method. Figure 92 shows a possible approach for such an instrument: If necessary, the incoming signal is either amplified or attenuated through an initial low-noise amplifier. Second we have a highly stable reference frequency source ("highly stable" meaning minimum phase jitter and voltage noise). This source can either be integrated into the instrument, or be supplied by the user, and its frequency must match the average (carrier) frequency of the signal to be measured. Among the lowest-jitter sources commercially available are oven-stabilized crystal oscillators, but they are only tunable over a very limited frequency range (maybe 1 kHz for a 100 MHz oscillator), they are only available for a discrete set of frequencies, and the maximum attainable frequency is limited to around 100 MHz (keep in mind that at its base there is still a mechanical oscillation!). This means if we have to perform measurements on a signal with a frequency outside those restrictions we will have to make do with a more flexible source (though more noisy as well), e.g. some high-quality tunable RF generator.

The measurement itself is a relative measurement of the phase jitter of the signal under test (frequency f_{signal}) to the phase jitter of the reference source (frequency f_{ref}). Both signals are fed into a mixer, which is a device producing an output signal proportional to the *product* of the two input signals. The frequency spectrum of the output signal then contains the *beat frequencies* $(f_{meas} + f_{ref})$ and $(f_{meas} - f_{ref})$. A low-pass filter behind the mixer is sized in a way that it lets only the difference frequency $(f_{meas} - f_{ref})$ through. A feedback loop with long time constant looks at (and seeks to minimize) the time-averaged amplitude of this signal and so keeps measured signal and reference signal *in quadrature* by adjusting the reference frequency (or, in other words, it seeks to keep the reference frequency equal to the average frequency of the measured signal, and the *average* relative phase between them equal to zero).

If the measured signal has some phase noise (faster than the – slow – feedback loop time constant of a few Hertz), then this gets translated into amplitude noise on the mixer output. This noise signal is fed into a spectrum analyzer (can be relatively low bandwidth because it only looks at the *difference* frequency, and low bandwidth allows for ultra-low-noise designs) which yields a plot analogous to one side lobe of the fundamental in the basic spectrum analyzer technique (Figure 91), but with much better frequency accuracy and lower noise floor. The best-in-class instruments achieve -140 to -160 dB of noise floor compared to -100 to -120 dB for even the best spectrum analyzers.

Software allows the user to remove (or include) from the RMS jitter calculation any peaks in the spectrum caused by periodic jitter. Because only amplitude information is retained but the phase is lost, the method has no knowledge about the phase relationship between the different components, so it usually adds the periodic jitter calculated from the single peaks' area as RMS (root-mean-square), which is basically assuming they are independent from and uncorrelated to each other. This is a delicate assumption that will break down the very moment when we have non-sinusoidal jitter, because the Fourier components of this jitter will show up as separate – but strongly phase-correlated! – peaks.

11. DIGITAL SIGNAL PROCESSING

11.1 Averaging

Averaging is a very simple type of digital signal processing. Nevertheless it can not only reduce noise on the signal, it can also improve the system resolution beyond the raw digitizer performance.

11.1.1 Noise Reduction

Just like random jitter, random noise on a signal can be described by a Gaussian distribution, only now in the voltage axis (or whatever value we measure) instead of the timing axis. Again we consider as the "true" signal level to be the mean value of the distribution. And we already know what averaging over N repeats does to our distribution – it becomes narrower by a factor of $\sqrt{N}$. So e.g. in order to reduce the noise to $1/10^{\text{th}}$, we need to average over 100 repeats. Since the square root grows only very slowly – we have to increase the number of averages by a factor of 100 for a noise improvement by a factor of 10 – it clearly pays to start out with the least amount of noise possible.

11.1.2 Improving Timing and Amplitude Resolution

One additional – and very useful – property of averaging repetitive signals is that it can actually *increase* the available voltage and – indirectly – the timing resolution, as long as the signal carries some amount of noise. This sounds surprising at first, since the very goal of averaging was to reduce the effect of signal noise, so how does this come about?

Let's first assume a noise-free signal, e.g. of amplitude 150.5 (in units of voltage sample intervals) going into an 8-bit sampler (i.e. sample range 0 to 255). Since the sampler cannot sample fractional intervals, the measured value is always 150, which will not change even when we average over many acquisitions. Thus the limited sampling resolution creates an error of up to 0.5 sample intervals, which can be substantial for fast but low-resolution real-time samplers.

On the other hand now consider some random noise riding on the signal, of maybe ±1 sample interval size. For our particular example, the noise will make the signal sometimes fall into interval 151, sometimes into 149, but mostly into 150. If we average up sufficient samples, then the average will turn out to be very close to 150.5. So the sampler – which can really only resolve integer numbers at each sample event – is capable of providing fractions of an interval in resolution. This is equivalent to a higher-resolution sampler, and many oscilloscope data sheets indicate this somehow similar to "resolution: 8 bits (11 bits with averaging)".

Due to the statistical nature of noise, the signal-to-noise ratio improves with $\sqrt{N}$ for linear averaging (all samples weighed the same[132]) – remember the formula for the variation of the mean value! Thus the actual improvement in level accuracy when doubling the number of averages is 3 dB, or an equivalent of $\frac{1}{2}\log_2 N$ additional bits, e.g. 4 bits for averaging over 256 samples. Clearly we need a minimum of 1 bit of noise present for the improvement to work in all cases, which means a minimum loss of base accuracy of 1 bit to begin with, so the total effective resolution is at best

$$resolution(bits) = resolution_{sampler}(bits) - noise(bits) + \frac{1}{2} \times \log_2 N \quad (102)$$

This means we can obtain improvement even with more noise than necessary, but for the price of still more samples – the necessary number of samples increases exponentially. A very practical limit to this will be the long-term stability (drift) of both the scope and the signal as well as the available acquisition time.

Since the trace and its timing are reconstructed from the samples (potentially through use of some interpolation method), an improved voltage accuracy translates directly into improved timing measurement accuracy.

11.2 Interpolation

We have already dealt with this type of signal processing in section 7.4. Interpolation can greatly improve timing resolution for edge crossing measurements, especially on instruments with limited resolution, e.g. real-time scopes that would otherwise be limited to 25 ps timing resolution or worse. Nevertheless, we know that we should always beware of artifacts (preshoot, ringing) caused by the Gibbs phenomenon. This is a common theme when applying DSP to a signal – one often has to wonder if the

[132] Exponential averaging, where more recent acquisitions are weighed more strongly than older ones, still delivers a similar benefit, but not as much as linear averaging.

resulting data is "real" or some observed effects are artifacts of the DSP process.

11.3 System De-Embedding

As we have seen over the course of this book, any part of our setup will degrade the original ("real") signal to some extent – in time domain this means the waveform as recorded will be different – distorted – from the input waveform. In frequency domain we can see it as a frequency-dependent attenuation (or gain) and a frequency-dependent phase shift. We might suspect that if one knew exactly how the measurement setup degrades the signal, one could mathematically restore the original waveform from the recorded waveform. And in fact this is one of the applications of digital signal processing (DSP) in modern oscilloscopes and other measurement devices. With interpolation we have already seen one such application where one aims to improve the fidelity of the measurement through post-processing of the measurement data, but the achievable improvement using this method is limited and may even introduce artifacts in the waveform like overshoot and ringing. On the other hand such interpolation methods like sin(x)/x do not take into account any knowledge about the measurement system itself (other than the fact that it samples the signal at regular intervals), but rather attempts to extract the best possible information from just the measurement data.

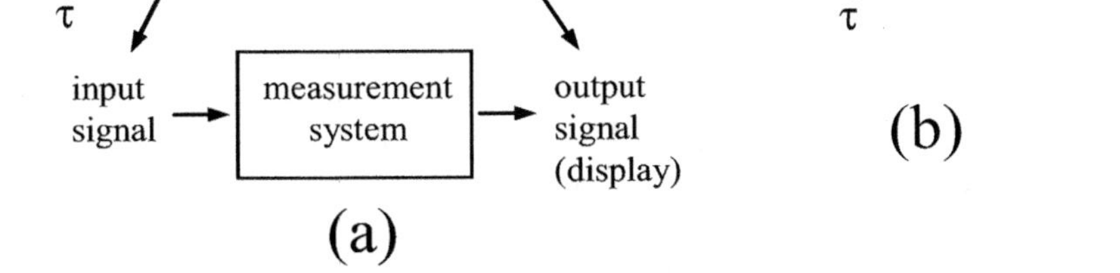

Figure 93: (a) Impulse response of a system, i.e. its output signal caused by an ideal infinitely short pulse of finite total power (a so-called Dirac delta function) as the input signal. (b) Decomposition of an arbitrary input signal into a sum of delta function shaped pulses.

The first step for further improvement is thus a detailed characterization of the measurement system's response to an incoming signal. We could for example put in a very narrow pulse – it has to be narrow (fast) because usually we will only see instrument limitations (distortion) when we get close to or exceed the instrument's bandwidth. In a theoretical treatment one employs a delta function as input, i.e. (casually speaking) a pulse that is infinitely narrow, infinitely high, but has a well defined total area (energy). Needless to say that in practice we can at best approximate such a step, not the least because the "infinite" signal amplitude would most likely kill our electronics. Figure 93(a) shows such a narrow pulse as well as the distorted, broadened measurement curve. The increased rise and fall time (and thus the broadening) is due to the limited bandwidth of the system. If the input pulse is much narrower than the resulting output pulse (i.e. the bandwidth of the input signal is much higher than the system bandwidth), then the output approximates the so-called "impulse response" *imp(t)* of the system.

For a linear system (and only for a linear system!) this impulse response is all we have to know to predict the output waveform for a given input waveform, or to re-create the input waveform based on the measured output waveform[133]: As a thought experiment we can always decompose any input waveform into a dense train of narrow delta function pulses, as shown in Figure 93(b). The input waveform is then identical to the sum of those pulses. The response (output signal) of a linear system for the sum of several input signals is the sum of the partial responses for each single input signal. Since the single input signals are (ideally infinitely) narrow pulses, the partial output waveforms are shaped like the (known!) impulse response, with amplitudes proportional to the area ("energy") of each particular pulse. In mathematical terms we say that the output waveform is the *convolution* of the input waveform with the impulse response, defined as an integral:

$$output(t) = input \otimes imp = \int_{-\infty}^{\infty} input(\tau) \times imp(t-\tau)\, d\tau \tag{103}$$

Unfortunately the integral makes practical numerical calculations for a given system rather difficult, not least because at least in theory one has to measure over a semi-infinite time interval. But things get much easier when one moves from the time domain into the frequency domain (by Fourier transformation of all three waveforms). In this case the convolution reduces to simple point-wise multiplication of each frequency bin:

[133] We should note once more that no real system is ever perfectly linear. Even the best oscilloscope, amplifier, cable etc. will exhibit some amount of nonlinearity that will limit the applicability of our theoretical considerations.

$$output(f) = input(f) \times imp(f)$$

$$\begin{aligned} input(f) &= FFT(input(t)) \\ output(f) &= FFT(output(t)) \\ imp(f) &= FFT(imp(t)) \end{aligned} \tag{104}$$

Note that all three items (input, output, and response) are complex (as opposed to real-valued) functions of the frequency. The absolute value of the response gives the attenuation (or gain) at each frequency, while the phase factor corresponds to a frequency dependent phase shift (delay).

The frequency-domain view gives another huge advantage of the time domain: In order to get the impulse response in time domain we had to generate an extremely narrow pulse, which poses many practical difficulties. On the other hand, in the frequency domain we can calculate the impulse response from any pair of input and output waveform simply as

$$imp(f) = \frac{output(f)}{input(f)}. \tag{105}$$

The only catch is that the frequency content of the input signal must not contain any zeros, i.e. frequency components with vanishing energy, because this will break the calculation of the division.[134] In practice it is already problematic when a frequency component becomes very small because then the unavoidable noise will dominate and produce arbitrary results. This is closely related to our earlier requirement in time domain that the input signal rise time must be fast against the system rise time, meaning there are input frequency components up to well above the system bandwidth which gives the frequency range of interest.

With known impulse response it is then just as easy to recreate the real input waveform based on the measured output:

$$input(f) = \frac{output(f)}{imp(f)}. \tag{106}$$

[134] The physical meaning of this is that if we do not send any signal components of a certain frequency into the system we cannot expect to be able to extract the response of the system to these components.

Again the denominator must not be zero at any point (or, in practice, it should not even become too small). Unfortunately there is no guarantee that a particular system will not contain such zeros in its response curve.[135]

In time domain (e.g. for an oscilloscope) the algorithm to improve the fidelity of the measured waveform is thus as follows:

1. Apply a calibration signal of known shape.
2. Record the output waveform
3. Fourier-transform both input and output waveform
4. Calculate the impulse response in frequency domain
5. Apply the signal to be measured and record the output waveform
6. Fourier-transform the output waveform
7. Calculate the input signal in the frequency domain based on impulse response and output signal
8. Perform an inverse FFT to obtain the calculated input waveform in time domain.
9. Repeat steps 5 – 8 with different input signals if necessary.

More often than not it will be difficult to know the exact waveform of the calibration signal, unless a reference instrument (with much higher performance than the "normal" instrument) is available. Fortunately there is a shortcut: If the input waveform is a simple fast-rising step (a step waveform is much easier to produce than a narrow pulse), with a rise time much shorter than the system's rise time, then the impulse response (in time domain) is simply the time-domain derivative of the output waveform produced by the step[136] (consequently called the "step response").

Another catch that is not directly visible from above considerations is even more serious: The necessary mathematical operations (FFT, deconvolution) are extremely sensitive to both noise and nonlinearities (we assumed

[135] In fact, we already know that e.g. a digital signal at a given constant data rate is *guaranteed* to have zeros at all multiples of the data rate.

[136] This is because we are assuming a linear system, a delta-function pulse is the derivative of a step function, and differentiation is a linear operation. In fact, differentiation and integration are shorthand notations for subtraction and addition of (infinitesimal) quantities, and we know that for linear systems we can exchange the response to the sum with the sum of the responses. Thus we can also exchange the response to the integral (the step function) with the integral of the responses (to the delta functions). Or, in the other direction, the response to the derivative of a signal is identical to the derivative of the response to this signal.

perfectly linear systems!) and tend to break down completely (i.e. produce entirely arbitrary results) the moment those effects exceed only a tiny fraction of the signal amplitude. In most cases random noise of just 0.1% (i.e. a signal-to-noise ratio of 40 dB) will make the results unusable.

Lots of effort is necessary to overcome those obstacles, since most oscilloscopes have trouble even hitting 40 dB SNR. Still it is worth keeping this method in mind. Averaging can improve the signal-to-noise ratio. Windowing and filtering can improve the quality of the result as well.

Some recent high-end real-time oscilloscopes have DSP of this type built in, but the achievable improvement in effective bandwidth is limited to about 20%. On the other hand above method is exactly what a vector network analyzer does (in time domain) when it gets calibrated, but here this is helped by the fact that such an instrument customarily exceeds 100 dB noise floor, and it is also more direct because all measurements as well as the results are in the frequency domain to begin with.

References and Further Reading

Signal Integrity, Transmission Lines, Electronics:

[1] Eric Bogatin, "Signal Integrity – Simplified", Prentice Hall, 2004, ISBN 0-13-066946-6

[2] Howard Johnson, Martin Graham "High-Speed Digital Design", Prentice Hall, 1993, ISBN 0-13-395724-1

[3] Howard Johnson, Martin Graham, "High-Speed Signal Propagation" Prentice Hall, 2003, ISBN 0-13-084408-X

[4] Stephen H. Hall, Garrett W. Hall, James A. McCall, "High-Speed Digital System Design" John Wiley & Sons, Inc., 2000, ISBN 0-471-36090-2

[5] Brian Young, "Digital Signal Integrity: Modeling and Simulation with Interconnects and Packages", Prentice Hall PTR, 2001, ISBN 0-13-028904-3

[6] Stephen G. Konsowski, Arden R. Helland, "Electronc Packaging of High Speed Circuitry", McGraw-Hill, 1997, ISBN 0-07-035970-0

[7] William R. Blood., Jr., "MECL System Design Handbook", Fourth Edition, Motorola Inc., 1988

[8] Paul Horowitz, Winfield Hill, "The Art of Electronics", Second Edition, Cambridge University Press, 1989

[9] Lee W. Ritchey, "Differential Signaling Doesn't Require Differential Impedance", Printed Circuit Design Magazine, March 1999

[10] Agilent Technologies, Inc., "Differential Trace Design Rules: Truth vs. Fiction", UltraCAD Design, Inc., and Mentor Graphics Corp., 2002

[11] Jiayuan Fang, Jin Zhao, "Low Impedance Power Delivery over Broad Frequencies", Printed Circuit Design & Manufacture, Sept. 2003

[12] Howard Johnson, "Diode Terminations", Online Newsletter Vol. 2, Issue 19, July 1998, at www.signalintegrity.com

[13] Howard Johnson, "Rise Time of Lossy Transmission Lines", EDN Magazine, Oct 2003

Oscilloscopes & Probes:

[14] Ian Hickman, "Oscilloscopes" (5th edition), Newnes, 2001, ISBN 0-7506-4757-4

[15] Ian Hickman, "Digital Storage Oscilloscopes", Newnes, 1997, ISBN 0-7506-2856-1

[16] "ABCs of Probes", Tektronix Application Note 60W-6053-8, Jan 2004

[17] "XYZ of Oscilloscopes", Tektronix Application Note 03W-8605-2, May 2001

[18] Nihal Kularatna, "Digital and Analogue Instrumentation", The Institution of Electrical Engineers, 2003, ISBN 0-85296-999-6

[19] Mark Burns, Gordon W. Roberts, "An Introduction to Mixed-Signal IC Test and Measurement", Oxford University Press, 2001, ISBN 0-19-514016-8

[20] Agilent Technologies, "Understanding Oscilloscope Frequency Response and Its Effect on Rise-Time Accuracy", Application Note 1420, 2002

[21] Dan Strassberg, "Getting the Most from Wideband Scopes – Dos and Don'ts", EDN Magazine, Feb 2005

[22] Richard Van Epps, Sheri McMasters, "High-Speed Probing for Superior Signal Fidelity", Tektronix Application Note

[23] Andy Frost, Don Whiteman, Jason Tsai, "Are You Measuring Your Circuit or Your Scope Probe", Agilent Technologies Applications Note, 2000

Statistics & Mathematics:

[24] W. J. DeCoursey, "Statistics and Probability for Engineering Applications", Newnes (Elsevier Science), 2003, ISBN 0-7506-7618-3

[25] Richard G. Lyons, "Understanding Digital Signal Processing", Second Edition, Pearson Education, 2004, ISBN 0-13-108989-7

[26] William H. Press, Saul A. Teukolsky, William T. Vetterling, Brian P. Flannery, "Numerical Recipes in C", Second Edition, Cambridge University Press, 1992

[27] Tektronix, "CSA/11000 Series Statistical Measurements", Technical Note 47W-7973, 1991

[28] James R. Andrews, "Deconvolution of System Impulse Responses and Time Domain Waveforms", Picosecond Pulse Labs Application Note AN-18, Nov. 2004

Noise:

[29] Joshua Israelsohn, "Noise 101", EDN Magazine, Jan 2004

[30] Dan Seslar, "Opamp Noise and Stability", Texas Instruments Presentation, 2004

[31] Chris Bowick, "RF Circuit Design", Appendix B: "Noise Calculations", Newnes, 1982, ISBN 0-7506-9946-9

Jitter Measurement:

[32] Osvaldo Buccafusca, "Back to Test & Measurement Basics: Jitter Measurements", Test & Measurement World, August 2004, download from www.tmworld.com/jitter-tutorial

[33] Bernd Laquai, "Spectral Error Density Analysis – A New Approach to Identify Jitter Sources in SOC Devices", Test & Measurement World, June 23, 2003, download from www.tmworld.com/SOCjitter

[34] Tom Waschura, "Using a BER Tester to Make the Fastest Eye Diagram, Mask Template, and Jitter Tests", Application Note SR-TN028, SyntheSys Reasarch, Inc., Jan 2003

[35] Andy Kuo, Touraj Farahmand, Roberto Rosales, Sassan Tabatabaei, Andre Ivanov, "Bounded Uncorrelated Jitter (BUJ) Characterization Platform for High-Speed Interconnects", IEEE Design and Test of Computers, 2005

[36] Sassan Tabatabaei, Michael Lee, Freddy Ben-Zeev, "Jitter Generation and Measurement for Test of Multi-Gbps Serial IO"

[37] T. Yamaguchi et al., "Effects of Deterministic Jitter in a Cable on Jitter Tolerance Measurements", International Test Conference, 2003, pp. 58-66

[38] "Measuring Jitter in Digital Systems", Agilent Technologies, Application Note 1448-1

[39] J. Wilstrup, "A Method of Serial Data Jitter Analysis Using One-Shot Time Interval Measurements", International Test Conference, 1998, pp. 819-823

[40] J. Quirk, "Measuring Deterministic Jitter with a Scope", Communication Systems Design, July 2002, pp. 26-33

[41] Slobodan Milijevic, "Jitter Generation and Measurement with Off-the-shelf Test Equipment", Zarlink Inc.

[42] "Jitter Analysis: Correlation: Measurement Techniques, Measurement Methods, Measurement Equipment, Mask Testing, Comparing Results", Tektronix Application Note, Jan 2003

[43] Art Porter, "Serial-Data Constraints Define Scope-Memory Depth", EDN Magazine, April 2004

[44] "Using the Architecture of the Tektronix 11801A Oscilloscope to Make Precise Jitter Measurements on Waveforms", Tektronics Inc. White Paper, 1998

[45] "Measuring Random Jitter on a Digital Sampling Oscilloscope", Maxim Application Note HFAN-04.5.1, Rev 0, Aug 2002

[46] John B. Rettig, Laszlo Dobos, "Picosecond Time Interval Measurements", IEEE Transactions on Instrumentation and Measurement, vol. 44, no. 2, April 1995

[47] Steven D. Draving, "How Accurate Are Your Jitter Measurements?", DesignCon 2003

[48] "Measure Error Rates Quickly and Accurately", Tektronix Technical Brief, 1998

[49] Mike McTigue, "Single-Ended or Differential? That Is the Question", EDN Magazine, Feb 2003

[50] Mike Tranchemontagne, "Making Timing Measurements on Differential Clock Signals", Tektronix Application Note, 1996

[51] Bradley J. Thompson, "Got the Time? An Introduction to Time Measurements", Test & Measurement World, Jan 1999

[52] "Jitter Analysis Techniques Using an Agilent Infiniium Oscilloscope", Agilent Product Note, Oct 2002

[53] Nelson Soo, "Jitter Measurement Techniques", Pericom Application Brief AB36, Nov 2000

[54] Richard A. Quinnell, "Resolving Jitter Measurement Variations", Test & Measurement World, Jan 2004

[55] Martin Rowe, "Measure Jitter Three Ways: BER Testers, Jitter Analyzers, and Oscilloscopes All Play Key Roles in Communications Testing", Test & Measurement World, Jan 2003

[56] Steve Fraash, Russ Cook, Mike Li, "Use of Compliant Clock Recovery in Serial Data Test Application", 2nd IEEE International GHz/Gbps Test Workshop, Nov. 2005

Jitter Analysis:

[57] "Understanding and Characterizing Jitter", Tektronix Application Note 55W-16146-0, Tektronix, Inc., October 2002

[58] "Jitter Analysis 101", Seminar Handouts, Wavecrest Corp., 2000

[59] Ransom Stephens, "The Rules of Jitter Analysis", http://www.analogzone.com/nett0927.pdf, 2004

[60] Ransom Stephens, "Analyzing Jitter at High Data Rates", IEEE Communications Magazine, vol. 42, no. 2, Feb. 2004

[61] Yoshitaka Takasaki, "Digital Transmission Design and Jitter Analysis", Artech House, 1991, ISBN 0-89006-503-9

[62] Andy Kuo, Touraj Farahmand, Nelson Ou, Sassan Tabatabaei, Andre Ivanov, "Jitter Models for the Design and Test of High-Speed (Gb/s) Serial Interconnects", IEEE Design and Test of Computers, vol. 21, no. 4, Jul-Aug 2004, pp. 302-313

[63] Y. Cai, S.A. Werner, G.J. Zhang, M.J. Olsen, R.D. Brink, "Jitter Testing for Multi-Gigabit Backplane SerDes – Techniques to Decompose and Combine Various Types of Jitter", International Test Conference, 2002, pp. 700-709

[64] "Jitter Analysis: The Dual-Dirac Model, RJ/DJ, and Q-Scale", White Paper, Agilent Technologies, Feb. 2005

[65] T. Yamaguchi, M. Soma, "Extraction of Peak-to-Peak and RMS Sinusoidal Jitter Using an Analytical Signal Method", IEEE VLSI Test Symposium, 2000, pp. 395-402

[66] M.P. Li, J. Wilstrup, R. Jessen, D., Petrich, "A New Method for Jitter Decomposition Through Its Distribution Tail Fitting", International Test Conference, 1999, pp. 788-794

[67] "Jitter Separation – 50 Mb/s to Over 40 Gb/s Using the Agilent 86100C Infiniium DCA-J", Agilent Technologies White Paper, 2004

[68] "An Introduction to Jitter in Communication Systems", Maxim Application Note 1916, March 2003

[69] Martin Miller, "Transition Density: What Does It Affect and Why Is It Explicitly Specified within the LeCroy SDA?", LeCroy Technical Brief, Jan 2005

[70] Martin Miller, "Six Tales of RJ and DJ or a Discussion of Which Measurement Approach Is Best Adapted to Specific Needs", LeCroy Technical Brief, 2005

[71] Martin Miller, "Total Jitter Measurement through the Extrapolation of Jitter Histograms", LeCroy Technical Brief, Jan 2005

[72] Martin Miller, "The Difference between Edge-to-Reference and Edge-to-Edge Jitter Analysis", LeCroy Technical Brief, Jan 2005

[73] Martin Miller, "A Simple Method for Estimation of True Jitter for Period, Skew, Half-Period and Cycle-Cycle Jitter", LeCroy Technical Brief, Jan 2005

[74] "Jitter Analysis: DJ Analysis; DJ and ISI Measurement Floor", Tektronix Application Note

[75] "Jitter Analysis: Accuracy, Resolution, Jitter Noise Floor, Delta Time Accuracy", Tektronix Application Note, Mar 2003

[76] "Jitter Analysis: Algorithms: Period Jitter, Cycle-to-Cycle Jitter, Frequency Jitter, Time Interval Error, General Measurements, Clock-Data Measurements, Calculating Statistics", Tektronix Application Note, Jan 2003

[77] "Understanding and Performing Precise Jitter Analysis", Tektronix Application Note 55W-13769-1, Feb 2004

[78] Aaron Schultz, "Random Jitter – What Is Really Going On?", EDTN France, Feb 2002

[79] Mike Story, "Timing Errors and Jitter", dCS Ltd., Oct. 1998, Issue 2

[80] "Analyzing Jitter Using a Spectrum Approach", Tektronix Application Note 55W-15631-0, March 2002

[81] "Delta-Time Accuracy: The Good, the Bad, and the Worst Case", Tektronix Application Note 55W-13565-0, Oct 2002

[82] "Comparing High-Precision Real-Time Jitter Analysis to Equivalent-Time Jitter Analysis", Amherst System Associates Application Note, Oct 2004

[83] Michael Fleischer-Reumann, "Integrated Evaluation of Jitter", Agilent Technologies Applications Note, 2000

[84] "Understanding Jitter", Wavecrest Application Note, 2001

[85] Dan Strassberg, "Eyeing Jitter – Shaking Out Why Signals Shake", EDN Magazine, May 2003

[86] Guy Foster, "Anatomy of an Eye Diagram – a Primer", Application Note SR-TN0033, SyntheSys Reasarch, Inc., Oct 2004

[87] "Histograms Simplify Jitter Analysis of Random Jitter", Agilent Application Note AN 1200-9, 1998 & 2000

[88] Guy Foster, "Look for Jitter in a Bathtub", Test & Measurement World, Dec 2003/Jan 2004

Frequency Domain Measurements:

[89] "Spectrum Analysis: Amplitude & Frequency Modulation", Hewlett Packard Application Note 150-1

[90] "Spectrum Analyzer Measurements and Noise", Agilent Application Note 1303, Feb 2003

[91] "Phase Noise to Jitter Conversion", Fordahl Application Note AN-02-3, www.fordahl.com

[92] "Jitter Measurement Using Integrated Phase Noise 1: Fundamentals", Toric Limited, 2003, www.toric.co.uk

[93] I. Zamek, "Crystal Oscillators' Jitter Measurements and its Estimation of Phase Noise", IEEE International Frequency Control Symposium , 2003

[94] J. McNeill, "A Simple Method for Relating Time- and Frequency Domain Measures of Oscillator Performance", Proceedings of the 2001 IEEE Southwest Symposium on Mixed Signal Design, Feb 2001

[95] R. Poore, "Phase Noise and Jitter", Agilent Application Note, 2001

[96] B. Drahlkis, "Calculate Oscillator Jitter by Using Phase-Noise Analysis", Microwave & RF, Jan & Feb 2001

[97] Aeroflex, "PN9000 Automated Phase Noise Measurement System Application Note #1", download from www.aeroflex.com, 2003

[98] Aeroflex, "PN9000 Automated Phase Noise Measurement System Application Note #2", download from www.aeroflex.com, 2003

[99] "Phase Noise Measurement Using the Wavecrest Digital Timing System", Wavecrest Corp. Application Note No. 130, 2001

[100] "Spectral Analysis: Analyzing a Signal Spectrum", Tektronix Application Note 55W-15429-0, Jan 2002

Index